THE GREAT POWER-LINE COVER-UP

THE GREAT POWER-LINE COVER-UP

How the Utilities and the Government Are Trying to Hide the Cancer Hazard Posed by Electromagnetic Fields

PAUL BRODEUR

LITTLE, BROWN AND COMPANY

BOSTON   NEW YORK   TORONTO   LONDON

First Edition

Some of this material originally appeared
in *The New Yorker* as "Annals of Radiation:
Calamity on Meadow Street," as a "Department
of Amplification," and as "Annals of
Radiation: The Cancer At Slater School."

Library of Congress Cataloging-in-Publication Data

Brodeur, Paul.
    The great power-line cover-up: how the utilities and the
government are trying to hide the cancer hazard posed by
electromagnetic fields / Paul Brodeur. — 1st ed.
        p.   cm.
    "Part of this book appeared in The New Yorker in a slightly
different form as 'Annals of radiation: calamity on Meadow Street,'
July 9, 1990" — Verso t.p.
    Includes index.
    ISBN 0-316-10909-6
    1. Electric lines — Health aspects.   2. Electromagnetic lines —
Health aspects.   I. Title.
RA569.3.B754   1993
363.18′9 — dc20   93-22332

10 9 8 7 6 5 4 3 2 1

RRD-VA

Published simultaneously in Canada by
Little, Brown & Company (Canada) Limited

Printed in the United States of America

ACKNOWLEDGMENTS

The author wishes to thank Charles Patrick Crow, Josselyn Simpson, and Elizabeth Pearson-Griffiths, of *The New Yorker,* for their valuable assistance in preparing the portions of this book that appeared in the magazine.

To my mother

CONTENTS

PART ONE

CALAMITY ON MEADOW STREET

CHAPTER ONE

Something Is
Very Wrong Here

ON A FRIDAY EVENING in mid-January of 1990, Edna and Robert Hemstock received a visit at their home, in Guilford, Connecticut — a town on Long Island Sound about fourteen miles east of New Haven — from their friends Loretta and Fred Nelson, who also live in Guilford. Edna Hemstock, a vivacious woman in her middle forties, who is Robert's second wife, works as an office manager for a manufacturing firm; Robert Hemstock, a red-haired man, of Irish ancestry, who was then forty-nine years old, is a free-lance consultant for machinery design and product development. Loretta Nelson, a slender, somber woman in her early forties, works at a nearby electronics plant; and Fred Nelson, an affable, wiry, gray-haired man who was then fifty-four, is an oil-burner serviceman. The two couples had become acquainted a year or so earlier, when the Nelsons' seventeen-year-old daughter, Joyce, and Charles Hemstock, Robert's twenty-year-old son by his first marriage, who had been living together in the Nelsons' house on Meadow Street, learned that Joyce was pregnant. This was a cause of some concern to both sets of prospective grandparents, because in 1982 Joyce, who is called Missy by her family and friends, had been found to be suffering from the extremely rare combination of glomerulonephritis — a disease of the kidney capillaries, which resulted in her losing thirty-five per cent of the function of both kidneys — and partial lipodystrophy, a disorder that has caused a severe loss of fat under the skin around her face, neck, shoulders, and chest. Because of her kidney condition, Joyce was carefully monitored during her pregnancy by physicians at Yale University's School of Medicine; they did not expect her to be able to carry the baby for more than six months. Happily, however, Joyce carried her baby for a full nine-month term, and gave birth on October 23, 1989, to a healthy seven-pound-three-ounce girl. Sitting at the Hemstocks' kitchen table nearly three months later, Loretta Nelson made a point of remarking on how lucky they were to be grandparents, considering the threat that Joyce's kidney ailment had posed to her pregnancy. She added, "And considering all the other illness there's been on Meadow Street,' which caused Bob Hemstock to sit up and take notice.

"That was the first time I'd heard anything about a lot of illness on Meadow Street," Hemstock said recently. "Edna and I had visited the Nelsons on several occasions while Missy was pregnant, and we had met their neighbor Suzanne Bullock, whose seventeen-year-old daughter, Melissa, had been operated on for brain cancer earlier that year, but neither of us knew about any other unusual incidence of disease down there. So I asked Loretta what she meant, and she proceeded to tell Edna and me a story that we found astonishing and disturbing."

To begin with, Loretta told her hosts that Joyce had been a healthy child of eight when the family moved to 36 Meadow Street from the neighboring town of Branford, in

1979, and that within a year she had become sickly and begun to lose weight. Loretta said that the doctors at the Yale School of Medicine had informed her and Fred that the causes of Joyce's glomerulonephritis and lipodystrophy were unknown but that her lipodystrophy might be the result of genetic vulnerability and of some type of environmental insult. She went on to say that she and Fred knew of no other case of either disease in either one's family. After reminding the Hemstocks that Melissa Bullock, who lived next door, at 28 Meadow, had developed brain cancer, she told them that Jonathan Walston III, who had lived at 36 Meadow as a child and young man, had developed a brain tumor in the late nineteen-seventies, when he was living at 48 Meadow — the house next door on the other side. She then said that Jonathan's father, Jonathan Walston, Jr., who was born at 36 Meadow and had lived there most of his life, had died there of brain cancer in 1975. She also said that a woman who had lived at 24 Meadow — three houses over — until she died of asthma, in 1989, had developed a brain tumor in the early nineteen-seventies.

Hemstock was flabbergasted by what he heard because Meadow Street, which runs north to south and terminates at a salt marsh lying between Guilford and Long Island Sound, is only about two hundred and fifty yards long and has only nine houses on it. He said, "Good God, Loretta, do you realize how odd it is to have that much brain cancer on one small street?"

Loretta replied that she and some of her neighbors were concerned about the situation but didn't know what to do about it. She also said that she and Fred had begun to suspect that the disease on Meadow Street might be caused by an electric substation that stood diagonally across the street from their house.

Over the weekend of January 13th and 14th, Hemstock mulled over what Loretta had told him, and the more he thought about the situation on Meadow Street the more concerned he became about it, and the more certain that he should try to find out what lay behind it. He looked up "nephritis" in his *Merck Manual of Diagnosis and Therapy* and learned that it was a kidney disease of various origins, one of which could be connected with exposure to mercury salts. That rang a bell with him, because he thought he had read somewhere that transformers were often coated with mercury. He then looked up "transformers" in an electrical-engineering textbook and learned that they are used to either step up or step down electric voltage.

"First thing Monday morning, I called up the Connecticut Light & Power Company, in New Haven," he recalls. "I asked for customer service and went through four or five people until I got a lady who was some kind of supervisor in the customer-service complaint department. I told her I was calling about a health problem on Meadow Street in Guilford. I told her there had been a lot of brain cancer there, and I asked her if she knew about it and if it might be connected with the company's substation. 'Absolutely not,' she said. 'There have been a lot of foolish studies that don't prove any connection whatsoever between electromagnetic fields and illnesses.' She went on to say that you could get more exposure to electromagnetic fields from a toaster or a coffee maker than from substations or power lines. I thought it was very strange how quickly she had brought up the subject of electromagnetic fields, only to deny that they could be a hazard. At that point, I knew almost nothing about electromagnetic fields, and I had certainly never heard that they might be connected with illness — yet here she was alerting me to the possibility even as she assured me that it didn't exist. I told her I was shocked that she had given me such a quick answer. I said I was calling about a serious matter and was looking for real answers, not rhetoric. With that, she got kind of huffy. 'Well, I'm sorry you feel that way,' she replied. 'I'm just telling you what our studies have shown.' This got my Irish up. I told her I was going to research the problem and find out the answer on my own. I also told her to take down my name. 'I'm Bob Hemstock,' I said, 'and you're not going to forget it.' "

That afternoon, Hemstock went to the library in Guilford and, with the help of a computer index, spent an hour or so looking up references to electromagnetic fields. He then went to the libraries in Branford and North Branford, to the west of Guilford, and did the same thing. By the end of the afternoon, he had made a list of more than a dozen articles, published in various magazines, medical journals, and newspapers around the nation, about the association of electromagnetic fields from power lines and other sources with cancer and other diseases. At the same time, he decided to investigate other possible causes of the disease on Meadow Street, such as chemicals being used at the Clinipad Corporation's factory, on High Street — about a quarter of a mile east of Meadow — which manufactures antiseptic medicinal pads. At about 5:30 P.M., he called Loretta Nelson, who had just returned home from work, and told her about his conversation with the supervisor at Connecticut Light & Power, his research at the three libraries, and his decision to investigate the chemicals being used by Clinipad. He then suggested that she get in touch with some of her neighbors on Meadow Street and hold a meeting toward the end of the week so that he could present his findings to them and learn how they might want to proceed.

The following morning, Hemstock called WTNH,

Channel 8, the ABC television station in New Haven. He said he wanted to speak to someone about a health problem on Meadow Street, and was asked to leave his name and telephone number. An hour or so later, he received a call from Becky Morris, the assignment editor at the station, to whom he related the essential elements of Loretta Nelson's story. Morris was interested — not the least of the reasons being that in December Channel 8 had broadcast a human-interest piece about Melissa Bullock's recovery from surgery for brain cancer and her return to Guilford High School, where she had been a star player on the girls' basketball team — but she appeared reluctant to cover the Meadow Street story without seeing medical records and talking to some of the doctors who had been involved. Hemstock told her about the articles he had come across, linking exposure to electromagnetic fields with brain cancer and other diseases, and his suspicion that the substation and power lines on Meadow Street might be related to the unusual occurrence of illness there. Morris said that she would consider doing a segment on the situation, and would call him back.

Later that morning, Hemstock drove to the hiring office of the Clinipad factory, and, in the course of inquiring about a job, asked for a catalogue of the company's products. A woman in the hiring office handed him an employment application but refused to give him the catalogue. Hemstock drove home, called the factory, and told a sales representative that he was thinking of putting together an industrial first-aid kit that the company might wish to manufacture. The sales rep agreed to compile a list of the company's products and their chemical components. The list included acetone, alcohol, iodine, and a dozen or so other chemicals, and Hemstock picked it up that afternoon.

In the evening, he looked them up in his *Merck;* although acetone was described as toxic, none of them were listed as actual or suspected carcinogens. He spent much of Wednesday, Thursday, and Friday at the Guilford library, looking up the toxicity of the chemicals and reading articles in various publications about the biological effects and disease-producing potential of the sixty-hertz alternating current electromagnetic fields given off by power lines. (Current for industrial and household use is known as sixty-hertz alternating current, because it flows back and forth, first in one direction and then in the other, sixty times a second.)

On Friday evening, a group of people joined Loretta and Fred Nelson at their home: among them were Jonathan Walston III, who is called Jack; his wife, Leah; Suzanne Bullock; and Bob Hemstock. After Hemstock introduced himself to Leah and Jack Walston, who had moved away from Meadow Street in 1983 and now lived a mile or so to the east, he told the group about his research.

"I started out by saying that there had been far too much illness on Meadow Street for us not to acknowledge the existence of a major health problem, and that I was concerned about the well-being of my son and granddaughter, and of everyone else who lived there," Hemstock recalls. "I showed them the list of chemicals used at Clinipad and said that, while I didn't want to rule them out as a possible cause of the problem, my research had so far provided no evidence that any of them could produce cancer, or that any Meadow Street residents had been unduly exposed to them. In connection with that, I pointed out that the town's drinking water, which is from deep artesian wells and is pumped into a holding tank several miles away, had been tested in September and found to be safe. I also talked at some length about the possibility that polychlorinated biphenyls, or PCBs — chemicals that were once widely used in transformers and are known to be toxic — might have leaked from the Meadow Street substation, or been dumped in a manner that might have exposed the residents. I then told them what I had learned about the cancer-producing potential of electromagnetic fields from power lines, and handed out copies of some of the articles I had read on the subject."

Included in the information that Hemstock gave his listeners was the fact that several epidemiological studies showed that children living near high-current electrical-distribution wires giving off relatively strong magnetic fields were developing cancer or dying of it at about twice the rate of children who lived in homes that were not near high-current wires. The first of these studies had been conducted in Denver, Colorado, by a pioneering epidemiologist named Nancy Wertheimer, and her colleague Ed Leeper, who is a physicist. They had published their findings in March of 1979 in the *American Journal of Epidemiology*, which is put out by the Johns Hopkins University School of Hygiene and Public Health in Baltimore, and is considered one of the foremost epidemiological journals in the world. During the nineteen-eighties, consultants hired by the Electric Power Research Institute (EPRI), an organization located in Palo Alto, California, that is sponsored by major utility companies across the nation, went to great lengths to try to discredit the findings of Wertheimer and Leeper. However, in 1986, the results of their study were confirmed by David Savitz, an epidemiologist then at the University of North Carolina's School of Public Health and some colleagues at the University of Colorado, who had conducted a thorough investigation under the auspices of the New York State Power Lines Project. Since then, the link between exposure to magnetic fields given off by power lines and the development of cancer in children had been demonstrated several times over.

Also included in Hemstock's presentation was the fact that a large number of epidemiological studies and surveys showed that men whose occupations expose them to electric and magnetic fields, such as electricians, electrical engineers, electric-utility workers, power-station operators and telephone and power linemen, were developing cancer — particularly brain cancer and leukemia — at a rate significantly higher than that of other workers. He specifically cited a study showing that electric-utility workers in East Texas were developing brain cancer thirteen times as often as expected. Hemstock went on to tell his audience about experimental studies showing that low-level electromagnetic fields, such as those given off by some power lines, could suppress the immune system, and thus suggesting how such fields might lead to the development of cancer, and about studies demonstrating that such fields could adversely affect the central nervous system, alter the chemistry of the brain in test animals, and cause birth defects and miscarriages.

"When I finished sharing what I had learned about the harmful effects of electromagnetic fields, I said that although I had not completed investigating the possibility of chemical contamination, my suspicions were very strongly directed toward the substation as the source of the disease on Meadow Street," Hemstock recalls. "I said that if that proved to be the case we would be taking on Connecticut Light & Power — a huge entity — and would be in for a long, hard battle. For that reason, I advised everyone to gather together the pertinent medical records, and I said that we ought to think about hiring a lawyer to represent us."

During and following Hemstock's presentation, there was a good deal of give-and-take between him and his lis-

teners. Leah and Jack Walston said they were convinced that
the substation was a factor in Jack's brain tumor and in his
father's brain cancer, and also in a tumor that had developed on the left leg of their daughter, Ann, when she was
thirteen years old and they were living at 48 Meadow — directly across the street from the substation. A number of
stories and anecdotes were told. Jack Walston remembered
that at the age of three Ann had been playing in front of the
house at 48 Meadow on a summer day when the substation
exploded and a Connecticut Light & Power engineer, who
had been working on one of the transformers, was severely
shocked and blown high into the air. He went on to say that
Ann had been credited with saving the man's life, because
she had run into the house to tell her mother, and Leah had
called the police and an ambulance. Fred Nelson remembered standing in his kitchen, holding a cup of coffee, one
morning, when a squirrel short-circuited one of the transformers, causing the substation to blow up and fill the
kitchen with a bright-red flash of light. He added that people living on Meadow Street had always had to reset their
electric clocks three or four times a week, because the re-
lays in the substation were forever being tripped, inter-
rupting the electric service. That observation reminded
Jack Walston that light bulbs had continually been burning
out when he and his family lived at 48 Meadow — that they
sometimes went through half a dozen bulbs a week. Loretta
Nelson pointed out that practically everyone on the street
suffered from severe headaches, and that the problem had
become so acute in the Brunelle family, who then lived at
48 Meadow, that Kevin and Mercedes Brunelle had recently
taken their nine-year-old son, Kevin, Jr., to the Yale University School of Medicine's Yale-New Haven Hospital for a
complete checkup and a brain scan.

During the week of January 22nd, Hemstock continued to visit the Guilford library and read up on the biological effects of low-level electromagnetic fields. He also received a number of telephone calls from Becky Morris, of Channel 8, who said she was still interested in doing a story about Meadow Street. At 9:00 A.M. on Thursday, January 25th, he telephoned Dr. John Brogdan, a general surgeon and the Guilford health officer. "I called Dr. Brogdan because I thought he should be alerted to the situation on Meadow Street, and because I thought he might be able to give me some advice," Hemstock recalls. "The first time I called, he hadn't come in yet, but when I telephoned again, shortly after ten, his secretary put me through. 'Dr. Brogdan,' I said, 'I'm Bob Hemstock. I'm calling about a health problem on Meadow Street.'

" 'Do you live on Meadow Street?' Dr. Brogdan asked.

" 'No, I don't,' I said.

" 'Then what business is it of yours?' he inquired.

"I was really taken aback by that question. For a second or two, I didn't know what to say. Finally, I said, 'I'm a member of the human race, Dr. Brogdan, and I'm interested in humanity. I thought doctors were, too.'

" 'Are you a doctor?' he asked.

" 'No, I'm not.' At that point he suggested that I gather up the medical records of the people on Meadow Street and bring them over to his office so that he could look at them and see if there was a problem. I told him I thought that that was an absurd idea. 'That's not my job,' I said. 'That's your job. You should be looking into the situation.'

" 'Listen, I've received a call from one of the neighbors over there that you're a troublemaker and don't even live in the neighborhood,' Dr. Brogdan replied.

"That really got my Irish up. I said, 'Well, my name is Bob

Hemstock, and you're not going to forget that name.' And I hung up."

The next day, Becky Morris called Hemstock to say that Channel 8 had decided to do a story on the Meadow Street cancer cluster for that night's eleven o'clock news broadcast, that she would arrive at the Meadow Street substation at four-fifteen, and that Kevin Hogan, a reporter, would interview him and some of the residents.

"Hogan asked a lot of questions and the camera crew shot a lot of tape," Hemstock recalls. "Most of it got left on the cutting-room floor, but some of it did make the Friday-night eleven o'clock news."

The news segment started out with Loretta Nelson standing in the yard beside her house and expressing concern about the health problems in the neighborhood. It then showed Fred Nelson also voicing worry, and he was followed by Hemstock, who stood in front of the substation and said it appeared that the substation was involved. Next, the camera turned to Kevin Brunelle, Jr., the nine-year-old boy who lives directly across the street from the substation, and he described his illnesses and the CAT scan he had been given at the Yale-New Haven Hospital, which had been negative. Then the program showed Jack Walston, talking about the brain tumor he had suffered. Finally, John Gustavsen, a spokesman for Northeast Utilities, the parent company of Connecticut Light & Power, gave the company's assessment of the situation. "It's a terribly unfortunate coincidence, and nobody to our knowledge has contacted our company, at least more than once," Gustavsen said. "If it's electromagnetic fields we are talking about, we are aware of studies that show a correlation statistically only. They don't demonstrate cause and effect."

On the morning of Monday, January 29th, yellow trucks

bearing the logo of Northeast Utilities were to be seen everywhere in Guilford. Three trucks, together with a supervisor's sedan, were parked on Meadow Street in front of the substation, and half a dozen men were working inside. Another yellow truck, carrying three workers, was parked beside a utility pole in front of a house at 140 Water Street, just around the corner from Meadow. Two of the workers were standing on the sidewalk by the pole, and the third worker had climbed the pole and was examining the wires strung from its crossbar. A Northeast Utilities sedan was parked farther east on Water Street, toward the center of town and a young man and a young woman who had got out of it were walking slowly along the sidewalk. The young man was calling out the serial numbers of distribution transformers that were mounted on utility poles, and the young woman, who was carrying a clipboard, was writing them down. Still farther along the street, a yellow van belonging to Northeast Utilities and bearing the sign "INFRA-RED SURVEY" was parked beneath the wires, and a worker wearing coveralls and carrying some kind of instrument in his hand was standing beside it. Hemstock, who was driving along Water Street at the time, parked on Meadow, got out, and walked back toward him.

"As I approached, he put the instrument in the back of the van, where there were a lot of gauges and equipment, and closed the door," Hemstock recalls. " 'What're you guys doing?' I asked, and he said, 'Oh, we're just looking for energy loss — heat loss — so we can improve the electrical service.' "

On Thursday, February 1st, the *New Haven Register* ran a story about the situation on Meadow Street, beneath a headline saying "POWER STATION STIRS FEARS OF HEALTH RISK." The story was by Judith Lyons, a staff writer for the newspaper, who had interviewed several of the residents. "Something is very wrong here," Kevin Brunelle told her. "My son and I are experiencing terrible headaches. Four people who have lived on this street have had brain tumors."

Brunelle's wife, Mercedes, told Lyons that brain scans and other tests had not been able to determine the reason for their son's headaches. "I'm scared," she said. "We have two other young children, too, to worry about."

Suzanne Bullock talked about her daughter Melissa's brain cancer, and said that the substation had not been ruled out as a possible cause. She also commented upon the unusual number of utility-company trucks that had been seen at the substation during the previous week. "We are glad they are there," she said. "We just want to see something done."

Henry Prescott, a spokesman for the utility, denied that any special measures were being taken in response to the residents' complaints. He said that workers had been sent to the Meadow Street substation to install devices that minimize short circuits.

Two residents of the street expressed skepticism that the substation was the cause of any illness. So did Robert Adair, a physicist at Yale University, who was described as an expert on electromagnetic energy. Adair had been quoted in the November 15, 1989, issue of the *Journal of the National Cancer Institute* that anyone who believed that electromagnetic fields could promote cancer "would believe in perpetual motion or cold fusion." Not surprisingly, he supported the position of the Connecticut Light & Power Company that there was no evidence that the substation had contributed to any illness. He dismissed a much publicized recent study conducted by Genevieve Matanoski, an epidemiologist at Johns Hopkins University's School of Hygiene and Public Health, and two colleagues; they had studied New York Telephone Company cable splicers, who are exposed to electric and magnetic fields from power lines, and found a significantly increased incidence of cancer — including seven times the rate of leukemia and almost twice the rate of brain cancer that was found among the company's office workers. Adair told Lyons that the electromagnetic fields produced by substations were "extremely weak," and that the currents going into and out of substations tended to cancel each other out.

The fact is that substations are known to produce very strong electromagnetic fields, and also transient high-frequency electromagnetic pulses. Moreover, because the currents carried by high-voltage lines going into a substation and those carried by lower-voltage distribution wires leaving a substation are almost never in balance, the magnetic fields they produce — invisible lines of force that readily penetrate virtually anything that happens to stand in their way, including the human body — almost never cancel each other out. They certainly did not at the Meadow Street substation, where measurements taken at various places near the peripheral fence of the facility shortly after Adair made his statement showed magnetic fields ranging from twenty to several hundred milligauss. (A gauss is a unit of measure for magnetic-field strength; a milligauss is one-thousandth of a gauss; and levels of between two and a half and four and a half milligauss have been associated in several epidemiological studies, including the Matanoski study, with the development of cancer in human beings.) Adair, however, declared that a television set would have a greater biological effect on Meadow Street families than the substation would. Yet magnetic-field levels given off by the average television set fall off to biologically negligible strengths within two to three feet of the appliance, whereas strong magnetic fields can often be measured within a hundred feet of a distribution substation, such as the one on Meadow Street, and within the same distance of the high-current distribution wires leading from it. Adair concluded by telling Lyons that he found it "inconceivable" that the substation could have "anything to do" with illness among nearby residents.

On the same day that the *Register* ran its story about Meadow Street, several residents received telephone calls from Dr. Brogdan's secretary, to tell them that on February 5th town officials were planning to meet with officials of Connecticut Light & Power and discuss the health problems that had been reported. The meeting between officials of Guilford and Connecticut Light & Power was soon cancelled, however, and not rescheduled. Over the six weeks following the broadcast and the news story, the utility's yellow trucks and vans kept showing up at the Meadow Street substation with such frequency that some residents of the street began to jot down their license numbers. The residents not only observed power-company employees working on the equipment in the substation but also noticed that a number of them were carrying instruments similar in appearance to gaussmeters — devices used to measure the strength of a magnetic field. They concluded that the company was engaged in an effort to reduce the amount of power being handled by the substation. One indication that this may indeed have been the case came in March, when an official of the Connecticut Department of Public Utility Control, in New Britain, told Hemstock that the Connecticut Light & Power Company had recently informed him that the Meadow Street substation was then receiving only twenty-two thousand six hundred volts — a substantial reduction, considering the fact that the facility had previously been fed voltage from several hundred-and-fifteen-thousand volt lines. Another indication was that the loud and constant hum given off by the vibration of the laminated iron sheets of the substation's transformers — a sound most Meadow Street residents had grown so used to that they'd ceased to notice it — suddenly diminished.

CHAPTER TWO

How Were Any of Us to Know?

A SUBSTATION is an assemblage of equipment — circuit breakers, disconnecting switches, transformers, and the like — that is designed to change and regulate the voltage of electricity. At power-generating plants, alternators produce medium voltages — typically, twenty thousand volts, or twenty kilovolts. The transformers at generating plants raise these to higher voltages — typically, a hundred and fifteen thousand volts, or a hundred and fifteen kilovolts, or else two hundred and thirty thousand volts, or two hundred and thirty kilovolts — which are required to transmit electrical energy economically over long distances to cities, towns, and other load centers. At those centers, distribution substations step down the transmission voltages to lower voltages — typically, 13.8 kilovolts or twenty-seven kilovolts — which then carry electric current to neighborhoods. There pole-mounted transformers step down the voltages further, to the two-hundred-and-forty-volt and hundred-and-twenty-volt levels required to operate household lights and appliances. In this respect, electric current — a flow of charged particles which always produces an electromagnetic field — can be likened to water flowing in a pipe, and voltage can be thought of as the pressure that pushes current through a circuit. Current for industrial and household use is known as alternating current because it is generated and supplied at a frequency of sixty hertz (formerly called cycles per second), which means that it flows back and forth — first in one direction and then in the other — sixty times a second. This man-made sixty-hertz frequency lies within the extremely-low-frequency (ELF) range, below the very-low-frequency (VLF) radio range, and it is entirely different from the earth's static magnetic field, to which the human body has been exposed during its entire evolutionary period. Indeed, every molecule in the brain and body of a human being standing in a strong alternating-current power-frequency magnetic field, such as is given off by any high-voltage or high-current power line, will vibrate to and fro sixty times a second — a phenomenon called entrainment, which, in turn, has been shown to alter the normal activation of enzyme and cellular immune responses in ways consistent with the promotion of cancer.

Like other distribution substations, the one on Meadow Street, which was built in 1932, was fed high voltage by a transmission line from Branford, and it sent out strong current at lower voltages on primary wires for distribution to various load centers in the Guilford area. (Stepped-down voltage invariably results in increased current, and the stronger the current the stronger the magnetic field.) As the demand for electricity grew in the region, the substation was required to handle and distribute an increasing amount of energy, and during the late nineteen-fifties and again in the sixties it was enlarged to accommodate additional high-voltage lines brought in from Branford, until it had grown to more than twice its original size. By the spring of 1972, the demand for electricity in the rapidly growing shoreline communities of Connecticut was such that Connecticut Light & Power proposed to run a hundred-and-fifteen-thousand-volt transmission line from Branford, through the northern sections of North Branford, Guilford, Madison, Clinton, and Westbrook, to Old Saybrook — some fifteen miles to the east — and to construct a bulk-supply substation in Madison for distributing electricity to load centers throughout the region. At the time, company officials estimated that a second substation would have to be built in Guilford in 1978, and another hundred-and-fifteen-thousand-volt line would have to be brought to the proposed substation by 1985 to supply the growing loads expected in the town. They told officials and residents of the affected towns that burying the proposed transmission line would cost at least six times as much as stringing it above ground.

Strong opposition from citizens' groups and zoning boards tied up the power company's proposal in regulatory proceedings for several years, and it was not finally approved by the Connecticut State Power Facility Evaluation Council until 1979. Meanwhile, the Meadow Street substation, which started out as a relatively small distribution facility, had begun to serve as a bulk-power substation for most of Guilford and for several other shore towns to the east. One set of high-current distribution wires crossed Meadow Street between the house at 48 and the house at 56, and traversed an adjacent salt marsh in an easterly direction, toward Madison. A second set of distribution wires ran north from the substation on telephone poles along Meadow Street, crossed Water Street, and continued north up River Street to the Boston Post Road, about a mile away.

A third set of distribution wires ran over the salt marsh in a northwesterly direction to Water Street near the West River, and then east along Water Street toward the center of Guilford, passing beneath the second set of wires at the corner of Water and Meadow. (High-current distribution wires are thick, large-gauge wires that are attached to utility poles with large glass or porcelain insulators.) These three sets of distribution wires were carrying very strong current — enough to supply the electrical needs of a wide region — and by 1974 they were creating magnetic fields powerful enough to interfere with television reception in the neighborhood.

During 1975, Robert Bryden, who lives at 140 Water Street, began to observe drastic warping and blurring of his television picture, and to experience pain and swelling in his eyes; his wife was suffering from swelling of the face and numbness and nerve tingling in one of her arms. Bryden then wrote more than a dozen letters of complaint to Guilford and Connecticut officials; to the Federal Communications Commission, in Washington, D.C.; to the representative for his congressional district; and to Senator Abraham Ribicoff. As a result of requests from the F.C.C., whose officials clearly believed that power lines were creating the interference, Connecticut Light & Power on several occasions sent representatives to Bryden's neighborhood to investigate the problem. However, after trying to correct the interference by working on the power lines, the company decided that its equipment was not to blame, and recommended that the problem be solved by the F.C.C., which informed Bryden that it had more important matters to deal with. Early in 1976, Bryden gave up and arranged to have cable television installed in his home. Many of his neighbors had already done so. Apparently, neither he nor they were aware that, just as someone can distort the lines of a drawing by jiggling the elbow of the artist who is making it, strong magnetic fields given off by power lines can distort a television picture by interfering with the path of the electron beam that forms the picture on the screen. Nor were they aware that in 1973 the members of a seven-scientist committee convened by the United States Navy had found the results of several Navy-financed studies of the biological effects of extra-low-frequency electromagnetic fields in human beings and animals so disturbing that they had recommended unanimously that the Navy (which suppressed their recommendation) warn a Presidential advisory panel of possible danger "to the large population at risk in the United States who are exposed to 60 hz fields from power lines and other 60 hz sources."

Meanwhile, in September of 1975, Jonathan Walston, Jr. (whose physician was Dr. Brogdan), died of brain cancer, at the age of fifty-four, at his home, at 36 Meadow. He had lived there from 1920, when he was born, until 1939, when he married a young woman named Marian Peck, and moved to North Street, about a mile away. His father, Jonathan, Sr., died, of a bleeding ulcer, in 1947, at 36 Meadow, and, shortly thereafter, Jonathan, Jr., Marian, and their three children — Jonathan III (Jack), Emily, and Wanda — moved there to live with his mother. During the next dozen years, Jonathan, Jr., worked as an iceman, as a firewood carter, and in construction; in 1960, when he was forty years old, he turned to lobster fishing. A few years later, he bought the house next door at 48 Meadow for twelve thousand dollars from a builder who had converted it into upstairs and downstairs apartments during the late nineteen-fifties. Jonathan, Jr., rented out both apartments at 48 Meadow and continued to live at 36 Meadow, and in May of 1971 he rented the downstairs apartment at 48 to his son, Jack, who moved there with his family.

Jack was born in 1942, while his father and mother were living up on North Street, and he went to live at 36 Meadow in 1947, after his grandfather's death. He quit high school at sixteen to work as a laborer, laying water mains in Guilford, and later went lobstering with his father. He lived at 36 Meadow until 1968, when he married Leah Craven, who was working as a secretary-receptionist for Guilford Pediatrics. During the first three years of their marriage, she and Jack lived up on Church Street, near the center of town, and their daughter, Ann, was born there in 1969. A few weeks after they moved to 48 Meadow, Leah gave birth to their son, Jonathan IV.

The year Ann was born, Jack's twenty-five-year-old sister, Wanda, who had recently married and was living in Branford, developed a nonmalignant ovarian tumor; it was removed by surgery. She was born on North Street in 1944, and had lived at 36 Meadow for almost twenty-three years. She was the first of four members of the Walston family to develop a tumor, either malignant or nonmalignant, while or after living at 36 Meadow or 48 Meadow. The second was her father, Jonathan, Jr.; the third was her brother, Jack; and the fourth was Jack's daughter, Ann.

Jack Walston is a heavyset man with dark eyes, dark hair falling over his brow, long dark sideburns that give him a rakish look, and an impish sense of humor. In the middle of his forehead is a tablespoon-size depression extending up to his hairline, where surgeons at Yale-New Haven Hospital removed a tumor from his olfactory groove, in the front part of the brain. In 1972, about a year after he moved back to Meadow Street — he was then working as a mechanic and welder for the Guilford Septic Tank Company —

he experienced a blackout while driving to Branford for National Guard duty. A year or so later, he blacked out again, in a bathroom at the National Guard armory. Early in 1979, he became irritable and began to experience terrible headaches and blurred vision — "It got so bad I couldn't see the welds," he says — so he went to a local eye doctor, who prescribed new glasses, which didn't do any good. In June, he went to another eye doctor in Guilford. This one told him that he had something growing on his optic nerve, and referred him to an eye doctor in nearby Middletown. The Middletown eye doctor said he had developed optic neuritis, and sent him to Dr. Thomas Walsh, a neurologist and ophthalmologist at Yale-New Haven Hospital.

"By time — September — things had got so bad that whenever I lay down on my side to watch TV I went blind on that side, whichever it was," Walston recalls. "I had also lost most of my ability to smell and taste. Dr. Walsh gave me a CAT scan that showed a large mass in the olfactory groove, and recommended that I go see Dr. Dennis Spencer, a neurosurgeon, who is the head of Neurosurgery at Yale-New Haven. After examining me, Dr. Spencer told me that I had a large tumor in the front part of my brain, and that I had to get it operated on as soon as possible, so I went into the hospital the following week, and on September 28th he operated on me for seven hours. He removed my forehead bone in two pieces and took out a meningioma — a generally nonmalignant but often fatal tumor — which was the size of a small grapefruit and had got entangled in my optic nerves and stretched them out as fine as ribbon. In order to take out the tumor, he had to sculpt out a small piece of my brain — about five centimeters' worth of the right frontal tip — which the tumor had been pushing against. Then he patched my upper sinuses with tissue taken from a muscle on my skull, put the two pieces of forehead bone back in place, and sewed me up. I got out of bed the next day, because they wanted me to get my lungs working, but some kind of fluid began running out of my nose. It turned out that the tissue patch had come loose and I was leaking brain fluid, so they did another CAT scan and found out I was getting air into my head when I inhaled. They put gauze under my nose and told me to lie flat, so my brain would heal, but my lungs weren't working right, and I developed double pneumonia and got put on a respirator and went into a semicoma for about a week and a half. When I came out of it, in early October, they set me up for another operation, during which they drilled a hole through my forehead bone and syringed out the air. However, they still had to replace the tissue patch, so I had to undergo still another seven-hour operation, in which Dr. Spencer took out my forehead bone all over again, and repatched my sinuses with more tissue, from my thigh."

Leah Walston is a calm, forthright woman. She says that since the operation Jack has been on Dilantin to prevent brain seizures, and that his short-term memory has been impaired. "If someone were to ask him his telephone number, he might give the number at 36 Meadow when he was a boy," she said not long ago. "He gets irritated when he can't remember recent dates and things."

Looking back on events, Leah realizes that there was an unusual series of health problems on Meadow Street all during the nineteen-seventies. "It's hard to know how many of them were related to the substation, which often blew up in those days, but, taken all together, they make you wonder," she said. "In 1972, I lost our third child because of a miscarriage. At about the same time, the woman who lived four houses over from us, at Number twenty-four, developed a meningioma just like the one Jack had later except that hers was attached to her skull bone, so she had to have part of the bone replaced with a steel plate. Then there was the time the substation blew up in the summer of that year. My sister, Brooke, was visiting us, and we were all over at Jack's parents' house, next door, in the evening, when a thunderstorm came up. I started back to our house with Ann, and Brooke and Jack came behind me with little Jonathan. While they were walking along the street, a bolt of lightning hit the substation. Brooke grabbed up little Jonathan, and she and Jack made a dash for the house, and while they were running there was a loud hum from the substation, and it got louder and louder until there was an explosion, as if a bomb had gone off. A huge flash lit up the neighborhood with all the colors of the rainbow — gorgeous blues, greens, and yellows — and hung over us for a second or two. The lightning bolt had put the substation out of commission and knocked the power out in the whole town.

"A month later, when Brooke was vacationing up on Cape Cod, she had a grand-mal seizure in the kitchen of the house where she was staying. She fell down and cut herself on the head, but she came straight home to Connecticut without seeing a doctor. Then, in 1973, she had two more seizures, so she did go to a doctor. He diagnosed the problem as epilepsy — an electrical disturbance in the brain — and put her on Dilantin. She stayed on Dilantin for a couple of years, and then weaned herself from it and never had any more seizures.

"Meanwhile, Jack's father had died of brain cancer, and when Jack developed his meningioma, in 1979, Brooke said there had to be something in the air on Meadow Street that was giving people cancer — first of all, Jack's sister Wanda, and then the woman at twenty-four, and then Jack's father, and now Jack himself. Brooke said we should look into the substation, because she felt there must be some connection, but we resisted the idea. About that time, we learned that little Jonathan had Osgood-Schlatter disease — a condition in which the knee tendons don't grow properly, and the kneecaps don't join with the tendons. Then, in 1982, Ann, who was thirteen, developed a nonmalignant tumor, on her left tibia, just below the knee. She had first started to have pain in the knee and leg that winter, but our local doctor said that that was normal for growing females, so we didn't pay much attention. In late August, however, a lump the size of a marble developed below her knee, and the skin around it turned brown. We took her to Dr. John Ogden, an orthopedic surgeon at the Yale School of Medicine, and he X-rayed her and made the diagnosis. When Dr. Ogden operated on her, in September, the tumor was embedded so deep in the shinbone that he had to go through almost to the last quarter inch. Ann wore a full hip-to-ankle cast for a month, and she was on crutches for another month. Brooke got angry when she learned what had happened, and insisted that we do something about the substation. She and I quarrelled about it until she finally gave up."

Leah continued, "In 1983, we moved from Meadow Street up to North Madison Road, near Guilford Lake. Around that time, we found out that Ann had developed a mild case of scoliosis, which is curvature of the spine. A year or so later, she developed blurred vision. Like little Jonathan, she had bad headaches practically the whole time we lived at 48 Meadow. We took her to Dr. Laura Ment, a pediatrician and neurologist at Yale-New Haven. She gave her an EEG and said Ann was suffering from temporal-lobe epilepsy. Dr. Ment put her on Dilantin, and although she had one epileptic seizure — in October of 1984 — she was able to graduate from high school in 1987 and then from the hair-dressing academy in Branford. In 1985, she developed a ganglion cyst in her left wrist. We took her to Dr. Kendrick E. Lee, an orthopedic surgeon at the Yale School of Medicine, and he removed it. The cyst has recurred three times, and it has now affected the nerves of her hand so badly that she can't work as a hairdresser. Back in 1980, I had developed some very large and incredibly painful keratinous cysts under my arms and in my groin. The doctors at Yale-New Haven said it was because my sweat glands were plugged up, so they operated on me and removed the cysts from under both my arms. Then, about the time Ann developed her ganglion cyst, my face swelled up terribly, and Dr. J. Cameron Kirchner, an ear-nose-throat surgeon at Yale-New Haven, had to remove the parotid gland from my left cheek, because it had become grossly inflamed and wasn't functioning properly. When you think of all the trouble we had — just one family — it makes you wonder."

Jack Walston said, "The builder who sold the house at 48 Meadow to my father had bought it from a couple whose son was born crippled and was never able to walk. They also had another son and a daughter, who were normal. The builder, who had three children, too, lived in the house only briefly, and moved away soon after he converted it into upstairs and downstairs apartments. All the bedrooms in both apartments faced the substation, just across the street. The upstairs apartment was rented to a couple, and they lived there for several years. They had three children. One was born with a deformed spine, and it was corrected by surgery at Yale-New Haven Hospital. After my father bought the house, he rented the upstairs apartment to a couple who had one son. The husband worked for a local company that manufactured electrical equipment for boats, and he had a steel plate in his head, because of a wound he had suffered during the Second World War. He later committed suicide with a gun. After that family moved away, my father rented the upstairs apartment to a couple who had a son who was born mentally retarded in 1967 and had epileptic-type seizures as he grew, and was treated at Yale-New Haven Hospital. My father rented the downstairs for several years to a couple who had two sons during that time. One was born with hypospadias — a defect of the penis — and the other was born with a hole in his heart. Both defects were repaired by surgery at Yale-New Haven. After those people moved away, another couple rented the apartment until, in 1971, Leah and I moved in."

"How were any of us to know that the substation might have caused all that trouble," Leah said. "One of the doctors who operated on Jack came to dinner at our house in 1980, and he didn't say anything to us about the substation being across the street. All the doctors who later treated Ann were told about Jack's brain tumor and the fact that his father had died of brain cancer, and none of them seemed to think there was anything unusual. Only my sister, Brooke, made the connection between the substation and what had happened to us, and we didn't believe her."

Jack Walston's mother, Marian, whose sixth child was born dead in 1954 because of a premature separation of the placenta, sold the house at 36 Meadow to a builder in 1976, soon after Jack's father died. The builder remodelled the house and rented it to a family. They lived there from 1977 until 1979. Then the builder sold the house to Loretta and Fred Nelson. Loretta and Fred have two children — Fred, Jr., who was born in 1968, and Joyce (Missy), who was born in

1972 — and they bought the house at 36 Meadow for thirty-seven thousand dollars in September of 1979, just at the time Jack Walston's brain tumor was diagnosed.

"About a year after we moved here, Joyce began to lose her face," Loretta recalls. "Her face was shrinking and hollowing out. In the spring of 1981, she developed a strep throat and a high fever, and her urine turned a dark-brown color. I was working at the electronics plant, so Fred took her to our pediatrician. He said she had a kidney problem, and sent us to Dr. Norman Siegel, a nephrologist at Yale-New Haven Hospital. Dr. Siegel performed a biopsy on her in May, and that's when he told us that she had glomerulonephritis and had lost thirty-five per cent of her kidney function. Other physicians at the hospital then diagnosed her lipodystrophy — the breakdown of fatty tissue that was causing her to lose her face. The combination of the two is extremely rare. In fact, Joyce was only the second case they had ever seen at Yale-New Haven. The doctors told us they didn't know what caused it, and we had no reason then to think that it might be associated with the substation here on Meadow Street, but now that we look back and think about all the cancer and headaches in this neighborhood we've begun to wonder. All of us — Fred, Sr., and Fred, Jr., and Missy and me — have had fierce headaches since we came to live here. A lot of other people on the street have them, too. The Brunelles, next door, for example, and a family who live up on the corner of Meadow and Water. And that's not all. Six months ago, Fred, Sr., developed a bump the size of a quarter on the back of his left hand, which got sore and caused him to lose the grip in his fingers. It turned out to be a keratoacanthoma, which is a benign tumor, and it had to be removed by a plastic surgeon. I'm afraid of the substation now. I keep remembering the couple who used to live next door to the big substation in Branford, in a frame house on Mill Plain Road, where the brick apartment buildings are now. The husband lived there in the middle nineteen-sixties, and when he was in his late twenties he died of cancer. It was all over his body. His wife also got cancer. I found out when I met her in the early nineteen-seventies at the hairdressing academy, while I was getting my hair cut. She told me she'd had chemotherapy. One of my biggest worries these days is Fred, Jr. He joined the Air-borne two years ago, when he was nineteen. He was stationed at Fort Bragg, North Carolina. Last year, when he was at Edwards Air Force Base, in California, they gave him an X-ray and told him he had either a tumor growing in his side or an extra rib. My other worry is little Amber, Missy's baby. When my friend Jane Harrison, who works with me at the electronics plant, heard that Melissa Bullock, next door, had brain cancer, she said we ought to get the baby off the street."

Loretta had become friends with Melissa's mother, Suzanne, when both were living in Branford during the nineteen-seventies. When Suzanne and her husband, Marshall, and their two children, Melissa and Corey, moved to Meadow Street, in the winter of 1979, Loretta came to visit them on several occasions. "I wish I had a house of my own," she used to say, and eventually Suzanne and Marshall persuaded her and Fred to buy the house next door.

Neither Melissa nor Corey encountered any unusual health problems while they were growing up on Meadow Street. By the time Melissa was sixteen, and a junior at Guilford High School, she had become an exceptionally beautiful young woman, almost six feet tall, who was a star player on the girls' basketball team. She was also a straight-A student, played the flute and piano, and had plans to go to college and ambitions to become a model. However, on the night of December 29, 1988 — a day before her seventeenth birthday — she suffered a grand-mal seizure on the high-school basketball court during a game against Plainville. She was taken by ambulance to the emergency ward at Yale-New Haven Hospital, and was given a CAT scan and examined by a neurologist on staff duty. He said that what he saw looked like "an old injury," and prescribed Tegretol to forestall any further seizures. She was back home in her bed in Guilford at one in the morning. The next day, Suzanne called her pediatrician, in Branford, and he referred her to Dr. Isaac Goodrich, a neurologist and surgeon in New Haven. On January 5, 1989, Dr. Goodrich sent Melissa to an outpatient radiology clinic for a magnetic-resonance-imaging examination. The M.R.I. showed a mass consistent with a low-grade tumor in the posterior lateral portion of the brain beside her left ear. Dr. Goodrich told Suzanne that it was a fairly common tumor, and was probably Grade One — that is, small — and nonmalignant, but that it should be removed as quickly as possible.

On January 19th, Dr. Goodrich operated on Melissa at the Hospital of St. Raphael, in New Haven, and removed the tumor, which was so close to the optic nerve controlling right-side vision that he warned Suzanne that Melissa's sight on that side might be impaired. He sent tissue samples from the tumor to the hospital's pathology laboratory for testing, and when the final result came back, a day later, he didn't believe it, so he sent a sample to Yale-New Haven Hospital for another opinion. The report that came back from Yale-New Haven was the same as the report from the first lab: Melissa had developed a Grade Three astrocytoma — advanced malignant tumor of the brain.

"Dr. Goodrich was real shook up when he told me the

news," Suzanne recalls. "He took it very badly. The tumor had become encapsulated, and that may have made the original CAT scan and subsequent M.R.I. diagnoses difficult. In any event, the cancer had spread far enough into the brain so that the prognosis for Melissa now had to be guarded. Late in January, Dr. Goodrich recommended that she begin radiation therapy, and starting in February she received two hundred rads in each of twenty treatments — five days a week for four weeks — at Yale-New Haven's Department of Therapeutic Radiology. She never got sick from the radiation, but she lost all her beautiful blond hair. Then, on March 27th, she went back to Yale-New Haven, and Dr. Joseph Piepmeier, a neurological surgeon, drilled two burr holes in her skull around the tumor site, inserted plastic tubing in each of them, and placed three seeds — tiny metal balls containing radioactive iodine — in each tube. The idea was that the iodine seeds would radiate outward and kill any cancer cells that remained in her brain. While she was in the hospital recovering from the operation, she suffered a full focal seizure, and they had to give her a huge dose of phenobarbital to pull her out of it. She also had trouble breathing and had to go on a respirator, but she was in top physical shape, because she's an athlete, so she bounced back quickly, and in two days she was able to come home.

"Dr. Goodrich had recommended that she undergo chemotherapy as well, and in June we went to Dr. Leonard Farber, an oncologist in New Haven. He gave her the first of two doses of an anticancer drug called CeeNU. She missed half a year of school, but, thanks to the homebound-studies program, she was able to start her senior year last fall. Since the operation, however, she has had trouble reading, and has gone from a Level One student — someone who gets straight A's — to Level Two. What happens is she'll get stuck on a word and lose the context of a sentence. She also has trouble understanding abstract concepts, such as irony. She often misuses words that have a similar sound — for example, 'pole' for 'pool.' In addition, she has lost part of her vision. In fact, from about one to four-thirty on the right side of both eyes she has a total blackout. She's back on the basketball team, though, playing first string and compensating amazingly well, considering her handicap.

"The strange thing about all this is that my mother, who's a retired registered nurse, never wanted me to move to Meadow Street. She was suspicious of living so close to a substation. Around the time Melissa's tumor was diagnosed, Loretta and I talked about the unusual number of cancer cases there had been on the street, so I asked Dr. Goodrich about it. I asked him if there was any chance that the substation was causing it and if the same thing could happen to Corey. In fact, I wanted Corey to have a CAT scan, but Dr. Goodrich said there was only a one-in-a-hundred-thousand chance of anything like that happening to him. Later, I read an article in a women's magazine about a study showing cancer in children who lived near high-current power lines. And recently I heard about a Vietnamese boy who got leukemia while living up on River Street, on the other side of Water, near the same high-current wires that come past our house here on Meadow. A few months ago, I developed a cyst in my breast, which my doctor tells me is going to have to be monitored with mammography every three months from now on. It seems there's no end to the disease around here, so if it isn't the substation and high-current wires that's causing it, what can it be?"

CHAPTER THREE

Going Through Light Bulbs Like Crazy

CANCER OF THE BRAIN is a rare disease in the United States, striking about one in every twenty thousand people each year. Nonmalignant meningioma is even rarer, occurring in about one in every hundred thousand people annually. The rate for brain cancer in Connecticut is approximately the same as for the rest of the nation, while meningioma occurs in about two to three in every hundred thousand Connecticut residents annually. Thus, to have two brain cancers and two meningiomas occurring in a span of twenty years among a handful of people living on a street of nine houses in Connecticut, or, for that matter, anywhere else in the United States, would be highly unusual — indeed, extraordinary — in and of itself. But the situation on Meadow Street is probably even worse than it appears, for the simple reason that when a thorough investigation of the health experience of its residents over the past thirty or forty years is conducted the cancer rate will undoubtedly have to be revised upward. Good reason for thinking so comes from the fact that during the spring of 1990 it was learned that still another resident of the street, Mrs. Judith Lehman Beauvais, had developed cancer. She was born in 1941, and she lived at 56 Meadow, within a few feet of the high-current wires that cross over from the substation, from the time she was seven until she was twenty-one. Then she married and went to live on Mulberry Point Road, in Guilford. In 1985, at the age of forty-four, she developed a malignant melanoma of the optic nerve, an extension of the brain, behind her left eye. It was treated with radiology at Yale-New Haven Hospital, leaving her partly blind in that eye and with limited vision in the other, and she died of it in July of 1990 after it had metastasized to other parts of her body.

The fact that two malignant brain tumors, a malignant eye tumor, and a nonmalignant brain tumor have developed in people living in four adjacent houses on Meadow Street which are situated across from the substation and close to its high-current wires, together with the fact that the other nonmalignant brain tumor occurred in a woman who lived just a few houses away, speaks for itself. The fact that one of the brain cancers, an astrocytoma, has been found in a seventeen-year-old girl makes the substation and its high-current wires additionally suspect, because astrocytoma can be expected to occur in only about one in every fifty thousand seventeen-year-old women each year. So does the fact that a nonmalignant ovarian tumor occurred in a twenty-five-year-old woman who had lived most of her life at 36 Meadow and that her niece — a thirteen-year-old girl, who had lived for eleven years at 48 Meadow — developed a nonmalignant tumor on her tibia.

Moreover, these tumors are simply some of the more serious afflictions that have beset the inhabitants of the dwellings at 28, 36, 48, and 56 Meadow over recent years. Ann Walston not only developed the tibial tumor at the age

of thirteen but suffered brain seizures when she was fifteen, and has since developed — and been operated on
for — painful and disabling ganglion cysts of the wrist. Her
mother, Leah, who had suffered a miscarriage while living
at 48 Meadow during the early nineteen-seventies, underwent surgery for painful and debilitating cysts during the
early nineteen-eighties, and, more recently, had to have an
inflamed parotid gland removed from her cheek. Leah's sister, Brooke, suffered brain seizures during the early
nineteen-seventies, when she was in her late twenties, and
shortly after visiting the house at a time when the substation was struck by lightning, causing it to blow up and send
powerful transient electromagnetic pulses throughout the
neighborhood.

Fred Nelson, who had lived at 36 Meadow for ten years,
developed a disabling growth on one hand, which had to
be removed by surgery. By the age of sixteen, his daughter,
Joyce, had developed glomerulonephritis and partial lipodystrophy — a kidney disease and a fatty-tissue disorder —
which are extremely rare in combination and may be the
result of some environmental insult. And Suzanne Bullock,
the mother of Melissa, has been found to have a suspicious-
looking cyst in one of her breasts.

Also disturbing is the extraordinary incidence of birth
defects in children who were conceived and carried at 48
Meadow. Serious birth defects occur in about one in every
twenty children who are born in the United States. However, no fewer than five of the nine children who were born
to parents living in this house during the ten-year span
from the middle nineteen-fifties to the middle sixties — it
was during that time that the substation across the street
was enlarged to handle higher and higher voltages — were
afflicted with serious congenital anomalies. Two other

children — Ann Walston and her brother, Jonathan Walston
IV — developed either spinal or ligamental disorders as
they grew. Additionally unsettling is the fact that a great
majority of the residents — both children and adults — of
36 and 48 Meadow during the past twenty years have experienced excruciating and recurring headaches.

During the previous thirty-five years, the fifty or so children and adults who were living in the four adjacent
houses on Meadow Street which are opposite the substation had been seen hundreds of times for their various ail-
ments by local physicians and pediatricians in Guilford and
Branford, and those doctors had sent most of the patients
whom they found to be suffering from serious illness, or
disease they could not identify, to the Yale University
School of Medicine's Yale-New Haven Hospital for examination, diagnosis, and treatment by a battery of oncologists,
neurologists, nephrologists, geneticists, anesthesiologists,
neurological surgeons, orthopedic surgeons, pathologists,
radiologists, and chemotherapists — specialists who, depending upon the individual case, proceeded to question
them about their symptoms and family medical histories; to
give them CAT-scan, X-ray, and EEG examinations; to delve
into their brains and innards; to excise their malignancies,
tumors, and other growths; to suture their wounds; to give
them postoperative radiology and chemotherapy; to pre-
scribe all manner of drugs and medicines; and to conduct
follow-up examinations, without asking any of these patients about the neighborhood in which the patient lived.
However, even if the cluster of malignant and nonmalignant
tumors of the brain among the residents of Meadow Street
had come to the attention of the doctors at the Yale School
of Medicine, or if local health officials had become curious
enough to stick pins denoting malignant neoplasms, non-

malignant tumors, and serious birth defects into a detailed street map of Guilford and thus been confronted by a forest of pins on Meadow Street near the substation, the physicians would no doubt have ascribed it to chance statistical variation — the rubric under which members of the nation's medical and scientific community have long chosen to file away (and avoid dealing with) cancer clusters. As for the particular cancer hazard posed by electromagnetic emanations from power lines and substations, the physicians are for the most part keeping silent until it appears safe to speak out — just as they did with the asbestos-cancer epidemic that will end up killing hundreds of thousands of shipyard and construction workers, and as they have done with the ongoing destruction of the ozone layer by chlorofluorocarbons, which, by significantly increasing the amount of harmful ultraviolet radiation reaching the earth, will inevitably result in a greatly increased incidence of malignant melanoma. In short, before physicians and public health officials feel free to acknowledge that substations and high-current power lines constitute a serious public-health hazard, it will undoubtedly be necessary to identify some additional Meadow Streets.

Unfortunately, this turns out to be not difficult to do. For example, in May of 1989, Dr. Sorrell Wolfson, a sixty-three-year-old physician, who was the director of the Salisbury Radiation Oncology Center, in Salisbury, North Carolina — a city of twenty-five thousand inhabitants about forty miles north of Charlotte, in Rowan County — had four brain-tumor patients referred to him for treatment in a single week. Wolfson, a native of Tampa, Florida, received his medical degree from the Vanderbilt University School of Medicine, in Nashville; did a two-year residency in pediatrics at the University of California Medical School in San

Francisco; and then had a year's fellowship in pediatric oncology and hematology at the Memorial Sloan-Kettering
Cancer Center, in New York City. After completing his train-
ing, he returned to Tampa and spent the next twenty-
two years treating children who had developed cancer and
various blood diseases. During that time, he was a
professor of pediatrics at the University of Florida, in
Gainesville, and at the University of South Florida, in
Tampa, where he became chairman of the Department of
Pediatrics.

Discouraged because he was seeing many of his young
patients die, Wolfson gave up his pediatric practice in 1980
to study radiation oncology, and since 1983 he has special-
ized in treating cancer with radiation therapy — first in
Tampa, and then, starting in October of 1988, in Salisbury.
Because he had treated only a few brain-tumor patients during the five years he practiced radiation oncology in Tampa,
and only one brain-tumor patient during his seven months
in Salisbury, he was naturally surprised to have four cases
referred to him in a single week. He became concerned
when he learned that all four patients lived near the small
town of China Grove, which is in the southwestern part of
Rowan County, about ten miles from Salisbury, and he became even more concerned when his interviews with them
revealed that six other people from that area had also developed brain tumors in recent years. At that point, Wolfson
called the Rowan County Health Department, and was referred to Dr. C. Gregory Smith, an epidemiologist with the
Environmental Epidemiology Branch of the North Carolina
Department of Human Resources, in Raleigh, and Dr. Smith
sent him a set of cancer-cluster-investigation forms to fill
out.

After verifying the ten brain-cancer cases in southwestern Rowan County through the medical records of hospitals in which they had been treated, Wolfson returned the cancer-cluster forms to Dr. Smith on June 21st. The ten cases went back to 1978, and all ten were primary brain cancers; that is, they had originated in the brain, and not metastasized there from elsewhere in the body. On July 5th, Smith wrote back to say that he planned to discuss the cases with an epidemiology intelligence officer from the Centers for Disease Control (C.D.C.), in Atlanta, who was scheduled to join the state's epidemiology branch on August 1st. Wolfson was elated to hear this, and two days later he gave a long interview to Rose Post, a reporter and columnist for the *Salisbury Post*. Post then wrote a two-part series on the cancer cluster, which appeared in the newspaper on July 9th and 10th.

In the first article, Post quoted Wolfson as saying that since the incidence of brain tumor was one in twenty-five thousand in the general population each year, four brain cancers were "a huge number for anybody to see in a week's period of time." Dr. Smith, for his part, told Post that the cluster "could be coincidence," and that the basic question to be answered was whether it had occurred by statistical chance or because of specific risk factors.

In the second article, Smith was even more cautious in his remarks to Post. He told her that his organization received reports of "perceived clusters" each week but did not investigate all of them. After noting that one out of every four people develops some form of cancer in his or her lifetime, Smith said that because there were so many retirement communities in North Carolina "you can go door to door and find lots of cancer." He went on to tell Post that his office was interested in the cluster that Wolfson had reported because it "involves a relatively rare type of

malignancy over a relatively short period of time." He said
that the state study would require statistics on brain cancer
in Rowan County and in the surrounding counties, but that
North Carolina's cancer-incidence registry was only two
years old, so "unfortunately, all we have in the way of cancer statistics at this time are death certificates." He assured
Post that the state investigators would look into the life
styles and occupations of the brain-cancer victims and also
into any family history of cancer.

On July 12th, a third article by Rose Post appeared. She
wrote that Dr. Wolfson had learned about twenty-four new
brain-tumor cases in Rowan County that week, including
another possible cluster of cases in the Trading Ford-
Dukeville area — two small communities about five miles
northeast of Salisbury, near the Duke Power Company's
Buck Steam Plant, on the Yadkin River. The next day, she
wrote that four additional cases — all of which, like the
twenty-four previous ones, were unconfirmed — had been
reported, bringing the total to twenty-eight. Dr. Smith said
he was not surprised by this. "That's not uncommon," he
told Post. "Many times that's what happens when a cancer
cluster is reported to the media." Smith went on to say
that the state would investigate and verify all the newly reported cases, and take special note of whether they were
primary brain cancers. "It's very common for many other
types of cancer to metastasize to the brain," he said. "Most
lay people don't realize that." Smith advised Post against
going out and interviewing any of the cancer victims herself, observing that untrained people asking questions
might provide patients with information that "becomes
part of their knowledge base and part of their answers,"
and "may be woven into their recollections of their own
life experiences." He also pointed out that cancer-cluster

reports "can turn into a firestorm pretty quickly," with the result that "everybody gets upset and makes urgent demands for things to be done, sometimes unreasonable demands."

After the publication of this fourth article, Post was unable to make further contact with Dr. Smith; he informed her that the epidemiology branch was receiving so many requests for interviews from television reporters, radio stations, and other segments of the news media that she would henceforth have to go through the Department of Human Resources' public-relations office. Meanwhile, other researchers joined him in expressing doubt that the pending investigation would turn up anything of much importance. "It's extremely rare that a study of what appears to be a cluster comes up with any likely cause," Leslie Boss, an epidemiologist at the Centers for Disease Control, told Pam Moore, another writer for the *Salisbury Post*. Boss went on to say that such studies "consume a tremendous amount of time and a tremendous amount of money," but that "from a political point of view" they had to be carried out. Boss noted that three or four people could be expected to die of brain cancer each year in Rowan County, but that "we don't get as excited about" the fact that there were thirty to thirty-five deaths from lung cancer and about thirteen deaths from breast cancer in the county each year. Dr. Glyn Caldwell, who said that he had participated in studies of more than a hundred cancer clusters during eighteen years he had spent at the C.D.C., agreed with the tone of Boss's remarks. He told Moore that he had never worked on a brain-cancer cluster, but that he was unaware of any such cluster in the United States which researchers had been able to authenticate. He went on to suggest that the new doctor in town (Wolfson) "may be the cause of the cluster" —

meaning that the cases could have existed for some time and it was only Wolfson's own discovery of them that made them appear significant.

On July 18th, the *Post* ran a story by Rose Post about the reported link between exposure of electric-utility workers to electromagnetic fields and the development of brain cancer and leukemia. On this subject she interviewed David Savitz, of the University of North Carolina, in Chapel Hill. He told her that he and some colleagues were conducting a study to determine whether power linemen and power-station operators were more likely to be afflicted with brain cancers and leukemia than other utility workers, such as office personnel and truck drivers, who do not have the same exposure to electromagnetic fields. (The study was being financed by the Electric Power Research Institute.) Savitz did not tell Post that just a few weeks earlier, at the annual meeting of the Society for Epidemiologic Research, he and a colleague had presented disturbing findings about deaths from brain cancer among electrical workers. He and his associate, who subsequently published their results in the *American Journal of Epidemiology*, had analyzed 410,651 deaths — among them 1,095 deaths from malignant brain tumors — occurring in sixteen states that participate in an industry-and-occupation-coding program of the National Center for Health Statistics, and had found that electrical workers had a fifty per cent greater rate of death from brain cancer than other workers.

In gathering material for her July 18th article, Post also interviewed Michael Mullen, a spokesman for the Duke Power Company, in nearby Charlotte. He told her that Duke Power was affiliated with EPRI, and that the weight of evidence of studies done during the past few decades "gives us confidence that we are providing electricity to our customers in a safe manner."

Post devoted a major portion of this article to the cluster of brain-cancer cases that Dr. Wolfson had reported from the Trading Ford-Dukeville area, near the Duke Power Company's Buck Steam Plant. That section read:

Two people who grew up as neighbors in the Dukeville community adjacent to the Buck Steam Plant have raised questions since members of both families died of brain cancer.

Bill Gilland's mother, Edna Gilland, died of primary brain cancer in January, 1988. "When she died," he said, "we started thinking of the people in that community who had died of the same thing."

He listed seven people. Two died 10 or more years ago, but the other five have died within the past three years.

Those names have been turned over to Dr. Sorrell Wolfson, director of the Salisbury Radiation Oncology Center. Wolfson prompted the study here when he contacted the Environmental Epidemiology Branch, State Division of Health Services, after he became concerned about four current brain-tumor cases in southwestern Rowan. Those four led to information about six others in the same area. All 10 cases have been verified.

OTHER CASES

The Post asked Duke Power about the study after questions were raised by relatives of people with brain cancer in China Grove and Trading Ford. The New York Times reported last week that accumulating scientific evidence had convinced many that there is cause for concern.

On that [the Dukeville] list is Walter Koone, who died of a brain tumor in March, 1986, right after his 50th birthday.

Koone's sister, Margaret Koone Murphy, who now lives

in Las Vegas, Nev., believes there's enough evidence of a connection to warrant further study.

"It's kind of amazing," she said, that so many people with brain tumors can be counted who worked at the power plant and lived in a community that had — at most — 300 people.

"I don't know what caused it," she said, "but I think there has to be something, whether it's food, water, or the electromagnetic field. I just feel like the numbers are too high." If the normal incidence of brain cancer is one in 25,000, "and you have that many brain cancers, not considering other types of cancer that people in this area have, it's a high number."

Mrs. Murphy and her two brothers grew up in Duke-ville village. Their father, who died Aug. 23 with cancer of the lung, worked at Buck Steam Plant all his life.

"And I'm home now," she said, "nursing my mother back from cancer of the breast."

Her brother, Walter, lived in the village until he graduated from high school. After four years in the Army, he moved to Belmont where he worked at the Allen Steam Plant until he died. Her other brother lives in Mount Holly and works for a Duke Power plant.

"My mother still lives on Dukeville Road," she said, "about a mile from the plant. Mrs. Gilland was my mother's closest friend."

Since health officials in North Carolina had been playing down the importance of the initial brain-cancer cluster, reported by Dr. Wolfson, it may not seem surprising that neither Dr. Smith nor anyone else at the state epidemiology branch apparently made any attempt to follow up the information about the incidence of brain cancers in the Trading Ford-Dukeville area which had been provided by Rose Post in the Salisbury newspaper. Somewhat more surprisingly, no other North Carolina newspaper, and no television or radio station, picked up Post's story. On January 6, 1990, however, the *Charlotte Observer* ran a long piece about the controversy over whether electromagnetic fields can cause cancer and other disease. The article quoted a review, prepared in 1989 for Congress's Office of Technology Assessment, stating that "the emerging evidence no longer allows us to categorically assert that there are no risks" entailed in exposure to electromagnetic fields, and that there was "some evidence to support the possibility that exposure can act as a cancer promoter." The *Observer* piece also cited the findings of the study recently conducted by Genevieve Matanoski, the epidemiologist at Johns Hopkins, who had found that the incidence of brain tumors among forty-five hundred New York Telephone Company cable splicers — men whose work near power lines and substations exposes them to electric and magnetic fields — was almost twice as high as that of the company's office workers, and that their leukemia rate was seven times as high. Nancy Wertheimer told the newspaper that a majority of the scientific studies conducted since 1979 showed that there was an association between exposure to electromagnetic fields and the development of cancer. However, Dr. Philip Cole, the chairman of the Department of Epidemiology of the University of Alabama, who had appeared at a 1987 congressional hearing to support the position of EPRI that power-line fields did not pose a health hazard, said that the findings of the studies were inconsistent, and added, "My own conclusion is that there's nothing going on here." David Savitz, for his part, declined an invitation from the newspaper to advise people to reduce their exposure to electromagnetic

fields. Doing nothing "may not be irrational," he told the
Observer. This seemed an unusually sanguine observation
from the author of a study that reported finding nearly
twice the expected number of cancer cases among children
living near high-current wires; and, as it happened, Savitz
was about to publish a new analysis of data from the same
study, showing four times the expected rate of brain cancer
in children whose mothers used electric blankets in their
first trimester of pregnancy, as well as higher than expected
levels of leukemia.

In any event, the members of the North Carolina Environmental Epidemiology Branch had been doing essen-
tially nothing about the cluster of brain cancers in Rowan
County which had been reported more than six months
earlier. Dr. Smith's plan to turn the cases over to the
epidemiology-intelligence officer from the C.D.C. in August
of 1989 apparently had not worked out, because in March
of 1990 — ten months after Dr. Wolfson reported the first
ten brain cancers in southwestern Rowan County, and eight
and a half months after Rose Post wrote about a second
brain-cancer cluster, in the Trading Ford-Dukeville area —
Dr. Peter Morris, a state epidemiologist, went to Salisbury
and spent two days interviewing the families and the sur-
viving victims in nine of the initial ten cases.

A week or so before Morris conducted these interviews,
he told Post that the state had already studied mortality
rates for Rowan and the surrounding counties, and that
"nothing out of the ordinary is going on in the area as a
whole, for brain cancer, at least." He said that he intended
to ask for detailed occupational and residential histories in
the nine cases, "to see if there is anything these people have
in common, or that a large proportion have in common."
He went on to say that it could be significant, for example,

if many or all of the brain-cancer victims had worked in the same plant twenty years ago. He also said that he did not expect the interviews to provide any definitive information, because cluster studies generally provide only suggestive information. During his stay in Salisbury, Morris did not visit Trading Ford or Dukeville, and when he was asked if he had reviewed the reported cases of brain cancer that Dr. Wolfson had sent to the epidemiology branch from that area he said he had not.

If Dr. Morris and his colleagues at the epidemiology branch had investigated the brain-cancer cluster reported among residents of the Trading Ford-Dukeville area, they would have discovered that all seven victims of the disease had indeed had something in common; they had worked at the Buck Steam Plant or had lived in Dukeville, a village of eighty-six houses, for married workers and their families, and a dormitory, for single workers, which the Duke Power Company had built near the plant, so that its employees could walk to work. Indeed, Dukeville was situated not only close to the plant but also adjacent to a large substation, where the medium voltages produced by the plant's alternators were raised to two hundred and thirty thousand volts and sent out over more than half a dozen transmission lines radiating from the plant, to carry electrical power throughout the central Piedmont.

The Buck Steam Plant — it was named for James Buchanan (Buck) Duke, who died in 1925 — was built in 1926. At that time, the company constructed forty-two four-, five-, and six-room houses for its employees, and those who occupied them paid no rent and were provided with free coal for heating. Twenty-two more houses were built in 1942, when Duke Power was called upon to produce extra electricity for the war effort, and another twenty-two were built in 1945, just after the war ended. In 1955, however, the company decided to sell its real estate in Dukeville, giving the resident employees the opportunity to buy their houses for between a hundred and a hundred and fifty dollars a room. The company offered the purchasers a number of lots on either side of Dukeville Road, at the top of a hill about a quarter of a mile away, for three hundred dollars apiece. About thirty of the houses were moved to the top of the hill, which is about midway between the power plant and Trading Ford, an old riverside supply post, and most of the other houses were moved elsewhere in the vicinity.

Today, almost nothing is left of the original village of Dukeville, but the Buck Steam Plant, which was partly shut down during the nineteen-fifties and sixties, is being renovated in preparation for being put back into operation. As for the three hundred or so people who lived in the village or worked at the plant (or did both), what is known of their health experience indicates that it has been unfortunate, to say the least. In addition to the seven people who have died of brain cancer, four others, who simply lived near the plant or the high-voltage transmission lines radiating from it, have also died of the disease. Moreover, a preliminary inquiry among people living on Dukeville Road reveals that there have been at least eight deaths from leukemia, lymphoma, and other cancers among people who had either lived in the village of Dukeville or worked at the Buck Steam Plant — including the death from lung cancer of a man who never smoked. By assigning this cancer cluster to the category of chance statistical variation, the North Carolina authorities seem to have overlooked an extraordinary hazard that — like the one on Meadow Street which confronted the physicians at Yale-New Haven Hospital for so many years — has been right in front of them. As for the likelihood that the inhabitants of Dukeville and the residents of Meadow Street were exposed to similarly powerful electromagnetic emanations from the substations near their homes, it is interesting to note that, just as Jack Walston remembers going through half a dozen light bulbs a week while he and his family were living at 48 Meadow, a former resident of Dukeville recently recalled "going through light bulbs down there like crazy."

Officials of the environmental-epidemiology section subsequently claimed that they had evaluated the seven cases of brain cancer, in order to "determine whether or not they should be included in our study of brain cancer in Rowan County from 1980 through 1989," and had found that "two of the seven cases had metastatic brain cancer, a different type of tumor originating in another part of the body and later spreading to the brain." They went on to say that four of the remaining cases were excluded from their study because the diagnoses of two of them were made prior to 1979, an unconfirmed diagnosis of another was made prior to 1979, and one of the victims lived outside Rowan County at the time of diagnosis.

In the final report of their study, which was entitled "Rowan County Brain Cancer Investigation," the North Carolina health officials stated that Rowan County did not have a significantly greater incidence of malignant brain cancer between 1980 and 1989 than each of the five surrounding counties. During a press conference at the Rowan County Health Department on October 25th, 1990, Dr. Morris told the *Salisbury Post* that brain cancer in the Trading Ford-Dukeville area during the ten-year period "was not studied as a separate cluster."

The rationale of the North Carolina health officials was faulty because they had not only failed to address the brain-

cancer situation in Trading Ford and Dukeville in its entirety but had also submerged the small part they did address in the larger study of Rowan County. In order to
understand how flawed their investigation had been, one
must remember that the power plant, which was built in
1926, was partly shut down during the nineteen-fifties and
sixties, and the eighty-six houses in Dukeville, which were
built between 1926 and 1945, were moved elsewhere in
1955. Thus, in addition to the one case of primary brain
cancer among Trading Ford and Dukeville residents that
the North Carolina officials included in their study, and the
four cases of brain cancer that they saw fit to exclude, other
people who were exposed to the electric and magnetic
fields from the plant, its substation, and its high-voltage
transmission lines by virtue of working at the plant or living
in the company village during the nineteen-thirties, forties,
and fifties may well have developed the disease and died of
it before 1979. By deciding not to include brain cancers diagnosed among residents of the Trading Ford-Dukeville
area before 1979, the North Carolina health officials decided not to investigate the health experience of people
who had worked at or lived near the Duke Power Company
plant — a decision that makes about as much epidemiological sense as a decision to study the incidence of gray
hair in a given population after excluding all those persons
in the study group who became gray more than ten years
earlier.

CHAPTER FOUR

Playing It Low-Key

I N APRIL of 1989, several weeks before the first brain-tumor cases were referred to Dr. Wolfson, a parent living in the neighborhood of Santa Rosa Lane, in Montecito, California — an affluent community of nine thousand people near Santa Barbara — got in touch with officials of the Santa Barbara County Department of Health Care Services to report a cluster of leukemias and lymphomas among children and young people in the area. When Dr. Alan Chovil, the director of Preventive Medical Services for the county, asked the California Department of Health Services, in Sacramento, to verify the report, the parent's concern was confirmed: six cases of leukemia and lymphoma — including one case of leukemia in a three-year-old child who had died earlier in the year — were diagnosed between 1981 and 1988 among residents of Montecito eighteen years of age or younger. According to state health officials, that was almost five times the number of leukemias and lymphomas that would normally be expected to occur during an eight-year period in a population the size of Montecito's, and was an event that could be expected to happen

by chance in only about two out of a thousand communities
of that size. After receiving the verification, Dr. Chovil asked
the Department of Health Services' Environmental Epidemiology and Toxicology Section to help investigate the
cluster, and plans were drawn up to interview the parents
of the afflicted youngsters and to conduct an environmental
investigation that would include reviewing historical records for information about possible contamination of the
area with pesticides or other chemicals, taking soil samples,
testing drinking water, and measuring the electromagnetic
fields at the local elementary school. The last factor was introduced into the study when Bronte Reynolds, the principal of the Montecito Union School, which is near the
intersection of San Ysidro Road and Santa Rosa Lane, told
state investigators that he and a group of parents were worried about electromagnetic fields because an electrical substation owned by the Southern California Edison Company
was situated next to the school's kindergarten playground,
and because an electric-transmission line crossed school
property.

During the first week of August, shortly before state investigators were scheduled to interview the parents of the
leukemia and lymphoma victims, the *Santa Barbara News-
Press* ran a story about the cluster, by a staff writer named
Melinda Burns. She reported that three of the afflicted children "live so close together" — presumably on or near
Santa Rosa Lane — "that they can see each other's homes,"
and that four of the children had attended the Montecito
Union School — a nearby elementary school with four
hundred pupils. It would soon be learned that five of the
children had attended Montecito Union; two of them had
developed leukemia, and three had developed lymphoma.
In her article, Burns interviewed a number of people about

the investigation that was in progress. Among them was Dr. Chovil, and he seemed ambivalent about it. "We've been trying to play it as low-key as possible," he said. "This is a tight little community, and we've been keeping in close touch with residents in the area." He also said, "There are more coincidences than you'd like to see," and added that he was "not a bit surprised that the citizens got concerned."

Dr. Lawrence Garfinkel, of the American Cancer Society, gave an assessment of the situation which sounded very much like the ones that the C.D.C. researchers had given Rose Post. "It's very, very rare to find a cause for a reported cluster," he said. "The conclusion that epidemiologists have to come to is that it's a chance phenomenon." Robert Schlag, a toxicologist with the California Department of Health Services, was even more pessimistic. "We'd like to find the cause of this," he said. "We don't think that we will." Schlag told Burns that by ruling out obvious environmental factors the state investigators would at least be able to assure the residents of Montecito that "there's no environmental hot spot here," so "they don't have to move out of town." He went on to say, however, that the cancer cluster among their children was "one of the sad facts of life," and that such a cluster "will happen again." Concerning the electromagnetic fields being given off by the substation and by the power line, Schlag said that very little was known about such radiation, and that the state was still in the process of obtaining the proper instruments to measure it.

On Sunday, September 24th — a time of the week when power demand is invariably lower than on weekdays, and when the strength of the magnetic fields given off by high-current wires is invariably reduced — staff members of the Environmental Epidemiology and Toxicology Section, who had never made electromagnetic-field measurements before, used borrowed equipment to measure the strength of the magnetic fields in a number of locations in the vicinity of the substation next to the Montecito Union School and a sixty-six-thousand-volt overhead high-current feeder line and three buried high-current lines that originate at the sub-station and pass forty or fifty feet in front of the school. (Curiously, no readings were taken inside the school.) Even on Sunday, however, a level of twelve milligauss was found under the power line opposite the substation, and one of four magnetic-field readings taken on the kindergarten patio was nearly two milligauss — a level just below that shown in three different epidemiological studies to be associated with twice the expected incidence of cancer among children.

The report of the state investigators, which was issued in draft form in December, under the title "Investigation of the Montecito Leukemia and Lymphoma Cluster," tried to put the best face on these measurements. "Exposure to electric and magnetic fields is an inevitable consequence of living in a society that uses electricity," it stated. "The earth has a DC [direct current] magnetic field of about 500 milligauss, and electric power lines and appliances generate AC EMF [alternating-current electromagnetic fields] of various magnitudes that decrease with distance from the source." With such language the authors of the report glossed over the fact that human evolution has occurred in the earth's steady-state, direct-current magnetic field, and ignored the fact that man-made sixty-hertz alternating-current magnetic fields will cause anything magnetic in their path, including the molecules of the human brain and body, to vibrate to and fro sixty times a second. The report went on to say that "we are all exposed to various degrees from ambient sources, household appliances, and from occupational sources," and to include a diagram of various sources of exposure which had been drawn up by EPRI.

The EPRI diagram indicated that magnetic fields close to household appliances are far stronger than those at the edge of a power-line right-of-way. It failed to point out, however, that the fields from almost all appliances fall off sharply within a few inches of their source, and that since people rarely stay as close as that to hair dryers, toasters, vacuum cleaners, and the like for eight hours a day, they cannot possibly be subjected to the same long-term chronic exposure as that of, say, a child attending a school situated within a few feet of a high-current feeder line. (Notable exceptions are electric blankets and electrically heated water beds.)

A page later, this paragraph appears:

The Southern California Edison substation and electric lines around the school are features relatively unique to Montecito Union Elementary School. Measurements were taken around school grounds to see if the substation was responsible for EMF's higher than those measured in other places in the area and in the country and to inform members of the community who were calling the principal and the health department about this concern. Measured levels fall within the range of measurements taken in other studies. They indicate that the EMF environment around the school does not appear to be different from other parts of the country even though there is a substation nearby.

After advancing the proposition that substations and the high-current distribution wires leading from them —

well known to electrical engineers as sources of powerful alternating-current magnetic fields because the wires are carrying all the current to be delivered to a given load area — do not contribute to the electromagnetic-field environment, the authors of the state report asserted that they had "found nothing that might have caused the cancers." They concluded by recommending that the California Tumor Registry "maintain the Montecito area under surveillance for cancer cases, and report yearly on trends in cancer occurrence in the area."

On December 14th, the day the findings of the state report were presented to more than a hundred Montecito residents in the school auditorium, *Montecito Life* — a weekly community newspaper — ran a front-page story, by a reporter named Laurie Koch Thrower, under the headline "RESEARCHERS CAN'T EXPLAIN MONTECITO'S CANCER CLUSTER." Dr. Chovil told Thrower that tests of Montecito's water showed it to be "squeaky clean," and that readings of a transformer in the schoolyard did not indicate that it was emitting any excess radiation. Richard Kreutzer, an epidemiologist with the Environmental Epidemiology and Toxicology Section, told Thrower that "nothing turned up in the investigation that explained the cases," and that "any combination of causes you can think of is possible." One possibility, of course, was that chronic exposure to sixty-hertz magnetic fields, which have been shown in experimental studies to be capable of hindering human T-lymphocyte cells from combating cancer, could have promoted cancer in children attending the Montecito Union School by suppressing their immune systems.

Many Montecito residents were puzzled by the lack of answers in the state report. Some were distressed by the way the California authorities had measured the magnetic fields near the school. One reason for the distress was that several parents had bought gaussmeters and taken their own readings of the magnetic fields on weekdays, when their children were in school. Not surprisingly, these readings turned out to be considerably higher than the ones taken by the state investigators on a weekend. In view of this disparity, Bronte Reynolds, the school principal, said that until more was known he was in favor of shielding the power lines. (Unfortunately, that would not help, because there is almost no way to shield the magnetic fields given off by power lines.) The state investigators responded by saying that they would address the concerns that had been raised, and present additional findings in their final report.

During January and February of 1990, Thrower continued to cover the story in detail for *Montecito Life*. On January 18th, she reported that Montecitans and their school representatives had been dissatisfied with the research conducted by county and state health officials, and had arranged for those officials and officials of Southern California Edison to remeasure the strength of the magnetic fields given off by the power lines near the school. John Britton, the area manager for Southern California Edison, told her that the company would cooperate fully in the new investigation. He pointed out, however, that state regulations mandating that new schools be situated at least two hundred feet from lines carrying a hundred thousand volts or more did not apply to power lines carrying less than that. Britton did not point out that a sixty-six-thousand-volt feeder line, such as the one passing within forty feet of the Montecito school, could at that distance give off magnetic fields exceeding those given off by high-voltage transmission lines at a distance of two hundred and fifty feet.

On February 8th — the day when funeral services were

held for an eighteen-year-old leukemia victim who had attended the school — *Montecito Life* ran a front-page story by Thrower under the headline "CANCER STRIKES SEVENTH CHILD IN MONTECITO." The new case involved a non-Hodgkin's lymphoma that had been diagnosed a few weeks earlier in a fifteen-year-old girl who had been attending the Howard School, which is a block and a half from Montecito Union. In spite of the fact that the Howard School sits within a few feet of the same sixty-six-thousand-volt high-current feeder line that passes in front of Montecito Union, county health officials told Thrower that they were not going to consider the case part of the cancer cluster under study, because they had found no evidence of a common denominator between it and the six original cancer cases. Dr. Chovil had another explanation. "We know we have too many cases," he said, but "if we haven't found it" — the cause — "among six, one more isn't going to help." He went on to say that so little was known about the effects of electromagnetic radiation upon health that when the results of the new measurements of the power-line fields at Montecito Union were made available he and his colleagues would not know what to do with them. "Even if we discovered all the children had been exposed to the same levels, it would not prove this was causing the cases," he said. Chovil then told Thrower that the seventh case of cancer had raised the hypothesis that "air travel could be a factor," because of the cosmic rays that air travellers are exposed to.

On March 1st, Chovil told Melinda Burns, of the *News-Press*, that the substation at the Montecito Union School was "completely irrelevant" to the ongoing study of the cancer cluster. (He was apparently unaware that the magnetic field given off by a high-current feeder line would be particularly strong along a portion of the line that was near a sub-station.) A day later, Jack Sahl, a research scientist working for Southern California Edison, who was monitoring the magnetic-field measurements being made at the school, echoed Chovil's opinion. "Montecito looks like just a normal school in terms of electromagnetic fields," he said. "These fields have been here as long as Montecito has had electricity. We're confident they're not associated with the cancer cluster." After conceding that he could not guarantee that there was no danger from magnetic fields, Sahl said that more studies were needed to verify the findings of the studies that had already been conducted. Perhaps the most interesting revelation came from Chovil: he told Burns that one student at Montecito Union (a second-grader, it was later learned) had been afflicted with testicular cancer but had not been included in the study, because the identified cluster did not include such cancers. This reasoning seemed arbitrary, in light of the fact that both Wertheimer and Savitz had found that deaths from cancer of all parts of the body were significantly elevated among children living near high-current wires. It seemed all the more arbitrary in light of the fact that the National Cancer Institute's Surveillance Epidemiology and End Results (SEER) data estimate the chances of a seven-or-eight-year-old child's developing testicular cancer to be nearly zero in one hundred thousand children per year. In any case, since cancer of any kind is a rare event in children, occurring annually in about one in ten thousand children per year under the age of fifteen, the appearance over an eight-year period of six cases of cancer in a school of four hundred students was worrisome, to say the least. The child-years at risk could be calculated at eight times four hundred students per year; that comes to thirty-two hundred child-years at risk. Six cases of

cancer out of thirty-two hundred child-years translates to
18.75 cases per ten thousand children per year. According
to the National Cancer Institute's SEER data, the all-sites
cancer rate for white children of both sexes, aged five to
nine, between 1983 and 1987 in the San Francisco-Oakland
area (the closest metropolitan area to Santa Barbara for
which SEER data exist) was 11.9 cases per hundred thousand children per year. Thus the cancer rate over those
eight years at the Montecito Union School — 18.75 cases
per ten thousand —was more than fifteen times the expected rate.

In an article that appeared in the *News-Press* on March
2nd, a staff writer named Pamela Harper reported that
many parents were alarmed by reports that the substation
and the high-current wires near the school might be producing dangerous magnetic fields, and that a mass exodus
of students to private schools in the area was feared. Harper
went on to note that Montecito had some of the most expensive homes in Santa Barbara County, and that "an un-
derlying fear that some parents — as well as residents in
the nearby neighborhood — have is that publicity about the
[cancer cluster] will bring property values crashing down."
And Barbara Koutnik, a real-estate agent and the mother of
two children who were attending Montecito Union, told
Harper that prospective buyers had recently taken several
houses out of escrow, and she suspected that they might
have used fear of cancer as a reason for terminating the
deals.

By now, there were a considerable number of people in
Montecito who wished to "keep a lid on the situation," as
one member of the Montecito School Board put it, and who
blamed the press for making disclosures about the cluster
which could adversely affect the community. On March 8th,

Montecito Life ran an editorial that addressed this issue
head on. It began:

> To report or not to report, that is the question. At least
> it sometimes is in small communities, where editors are
> occasionally rebuked for covering events that might be
> perceived by some as adversely affecting the public wel-
> fare.
>
> During a recent gathering of parents at the Montecito
> Union School, it was suggested, from the audience, that
> media coverage of the possible relationship between
> electromagnetic radiation levels emitted by power trans-
> mission lines and the high number of cancer cases in the
> Montecito Union School District was irresponsible. Panic
> potential was given as one reason. Unstated, but under-
> stood privately, is the effect this kind of news might have
> on property values. This attitude by no means reflects
> how most at that meeting felt. But it crops up frequently
> enough so that journalists must pause to consider why
> we relentlessly pursue the public's right to know.

The editorial concluded by declaring, "While we're not
sure how Montecito residents might react to adverse information concerning the effects of power line radiation, and
no case has been made at this time, our job is to provide
the conduit for conveying information from reliable
sources, even though we are ever mindful of the line from
Shakespeare's King Henry IV that 'The first bringer of un-
welcome news, hath but a losing office.' "

Meanwhile, on March 6th, Bronte Reynolds told a
hundred parents who had gathered in the school auditorium that after looking at the results of the latest
magnetic-field measurements he and a parents' task force
would make recommendations to the school board for

"immediate temporary measures" to restrict student exposure. Reynolds said that the measures might include roping off portions of the school, changing classroom seating, and calling for the rerouting of nearby power lines. Charles Cappel, a spokesman for an ad hoc parents' committee that had called for the meeting, said that, according to Southern California Edison, burying the wires, so that the magnetic fields would cancel each other out, might also provide a solution to the problem. "We were certainly not willing to let our children be made guinea pigs in some experiment," he told Thrower. "We wanted something done now — we weren't sure what we wanted it to be — but we wanted something done, by God!" According to an account of the meeting that appeared in the *News-Press*, a Montecito resident who asked whether the authorities planned to measure the magnetic fields under high-current wires on nearby streets was told by Dr. Chovil that there were no such plans, "because at the moment EMF is not considered to be a cause of cancer." Chovil was supported in this contention by Dr. George Fisher, a physician in Montecito, who said that there was no reason to be concerned about the school, because no laboratory studies had produced cancer in animals by exposing them to magnetic fields. "It makes just as much sense to rope off your toaster, your electric blanket, your computer and your TV set," Fisher declared. "I think there's a little hysteria going on." He was evidently unaware of Savitz' findings, or that the previous November — presumably as a result of the childhood-cancer studies and of other studies showing an association between electric-blanket use and miscarriages — *Consumer Reports* had advised children and pregnant women not to use electric blankets.

On March 15th, the parents' task force, in a meeting with

the Montecito School Board, recommended that a deter-
mination be made whether the metal cyclone fence along
the north side of the kindergarten playground was contrib-
uting to the electromagnetic field that had been measured
there, and that if this was found to be the case the fence be
replaced by one of wood. The task force also recommended that those portions of the kindergarten patio with
magnetic fields of more than two milligauss be roped off;
that benches and playground equipment on some terraces
beneath the sixty-six-thousand-volt feeder line be moved
farther away; that desks be removed from the southeast corner of a classroom that was near a high-voltage transformer
in one of the school's parking lots; and that stripes marking
ten-foot "restricted" zones be painted around the transformer and around a high-voltage circuit-breaker panel in
the fire lane outside the main building. At the same meeting, the parents' task force released the preliminary results
of magnetic-field measurements that had been made earlier
in the month by Enertech, an engineering consulting firm
in the Bay Area, and that had been paid for by Southern
California Edison. The results showed levels of between
four and six milligauss on the kindergarten patio and along
its fence; a level of almost seven milligauss on the benches
beneath the feeder line; a level of seventeen milligauss in
the corner of a classroom on the southeast side of the
school; and levels of between six hundred and a thousand
milligauss next to the transformer in the parking lot. Obviously, these measurements were far higher than the ones
taken on the Sunday in September by officials of the California Department of Health Services' Environmental Epidemiology and Toxicology Section.

At a meeting held at the school that night and attended
by about fifty parents, the school board voted unanimously

to accept the task-force recommendations. "To me these measures are like buckling your seat belt," one of the board members said. "What's the harm?" This question was addressed by Dr. Fisher, who said that he was representing twelve other physicians with children at the school. According to an account of the meeting written by Melinda Burns, Fisher claimed that the school board had no business making a decision based on the opinion of laymen, and added that the scientific evidence of a health hazard from exposure to electromagnetic radiation was inconsistent and contradictory, that the magnitude of risk was small, and that the focus on the school was misguided. "We believe that significant emotional trauma will come to our children if we selectively isolate areas of the campus as potential cancer zones," he declared. Dr. Abraham I. Potolsky, a hematologist in Santa Barbara, disagreed. "It's ridiculous not to take these minimum measures to avoid these potential hazards," he said.

During the last week of May, the California Department of Health Services issued a second draft report, in which it tried once again to put the best possible face on the situation at the Montecito Union School. The report's authors declared that magnetic-field levels at Montecito Union were "not unusually high in most parts of the school," and that there was no evidence that they posed a health hazard. They went on to say that the levels near the power lines along the north side of the school, which had been found to include some distribution wires buried in an alley along the kindergarten patio, were in the five-to-thirty-milligauss range, and they described these fields as "similar to what one is exposed to when near a common electrical household appliance such as a TV or a radio." This, however, had little, if any, relevance to the situation at Montecito Union, for the simple reason that one would be exposed to five milligauss from a television or a radio only if one sat within a few inches of it, and to thirty milligauss only if one pressed one's face to certain locations on its side.

On June 1st, officials from the Department of Health Services met with members of the Montecito School Board and the parents' task force to discuss the state's report. At the meeting, Charles Cappel pointed out that a person can decide how close to get to household appliances, but that attendance at school constitutes involuntary long-term exposure. Dr. Raymond Richard Neutra, the chief of the health department's special epidemiological studies program, declared that not enough was known about the effects of electromagnetic radiation to say if the levels measured at the school posed a health threat. According to Neutra, it would take at least two more years before enough data were collected to make such a judgment. Meanwhile, Southern California Edison informed school authorities that in the absence of conclusive evidence of a health hazard it would not underwrite the sixty thousand dollars necessary to unearth the wires near the kindergarten patio and move them farther away. On June 20th, the members of the Montecito School Board used similar reasoning in deciding not to underwrite the cost of moving either the underground wires or the sixty-six-thousand-volt overhead feeder line. They declared that not enough was known about the biological effects of electromagnetic radiation to warrant taking any permanent action. Six days earlier, the *News-Press* had reported something that Dr. Chovil had known since July of 1989, and that state health officials had known since December — that four cases of leukemia had developed in young children who attended the Montecito Union School in the late nineteen-fifties.

In a final report on the cancer cluster at Montecito Union, which was issued in December of 1990, Dr. Neutra and some colleagues at the Department of Health Services said that there was no indication of an unusual incidence of cancer at the school during the nineteen-sixties and nineteen-seventies, and no evidence that the strength of the magnetic fields given off by the power lines had increased sufficiently to account for the cancer cluster that had occurred there during the nineteen-eighties. They did not, however, investigate the health experience of the students who had attended Montecito Union during the nineteen-sixties and nineteen-seventies, nor did they ask Southern California Edison to furnish them with current-flow records that would have enabled them to know for certain whether or not the magnetic fields given off by the power lines had increased in recent years. As for the four cases of leukemia that were said to have occurred among children attending Montecito Union in the late nineteen-fifties, the state officials said that they did not possess sufficient data to make "a reliable assessment" of the matter. After noting that some epidemiological studies had found brain cancer to be associated with exposure to power-frequency magnetic fields, they said that there had not been an excess of brain cancer among residents of the town of Montecito. They made no mention of the fact that during the summer a teacher's aide with several years of experience in the kindergarten at Montecito Union — the portion of the school closest to the power lines — had developed a brain tumor. Moreover, in spite of the report that three of the cancer-afflicted children lived so close together that they could see each other's houses, they apparently made no attempt to ascertain whether these children lived in homes near high-current or high-voltage wires. In the concluding paragraph of their report, Neutra and his colleagues declared that because of "uncertainty" about whether power-line magnetic fields posed a health hazard, they were not prepared to make recommendations about safe levels of exposure. Instead, they suggested that members of the Montecito School Board and parents of Montecito Union children could use the information in the report "to decide if they wish to pursue any avoidance measures that they think may be indicated for their particular situation."

CHAPTER FIVE

A Growing Body of Evidence

MEANWHILE, students attending schools elsewhere in California and across the nation were being exposed to potentially hazardous magnetic fields given off by nearby power lines. Unfortunately, there were many such schools — one reason being that in years past utilities have often sold company-owned land cheap to cities and towns for schools and other municipal buildings, in return for assistance in acquiring easements for new power lines and substations; another being that, because of aesthetic considerations, land is less desirable in the vicinity of power lines than elsewhere, and can thus be purchased cheaply. For the latter reason, no fewer than five elementary schools in the Fountain Valley School District, which serves both Fountain Valley (a city of fifty-five thousand, in Orange County, about thirty miles south of Los Angeles) and part of Huntington Beach (a city of a hundred and seventy thousand, just to the west of Fountain Valley), have been built on land next to an easement for a pair of two-hundred-and-twenty-thousand-volt lines and a high-current feeder line. These lines carry power for Orange County from Southern

California's generating plant at Huntington Beach. As in the case of the Montecito Union School, high-current distribution wires have been buried in the ground beside many of these schools, and transformers have apparently been installed near some classrooms. As a result, magnetic-field strengths far exceeding those associated with the development of cancer in children and adults have been measured in areas immediately adjacent to several of these schools, including a level of twenty-five milligauss at the doorway of a boys' room at the Samuel E. Talbert Middle School; a level of fifteen milligauss at a kindergarten playground beside the Roch Courreges Elementary School; a level of nearly ten milligauss beside some outdoor lunch tables at the Harry C. Fulton Middle School; and a level of five milligauss in the kindergarten playground of the Tamura Elementary School. When this situation came to light in the spring of 1990, the superintendent of Fountain Valley School District said that the evidence of magnetic-field hazards was too inconclusive to warrant taking any action. Interestingly, in January, school-district officials had entered into a multi-million-dollar joint venture with a developer to build at least sixty homes on land occupied by the James O. Harper Elementary School, which had been shut down five years earlier and was next to the same power-line right-of-way as the other schools.

Not surprisingly, serious magnetic-field hazards existed elsewhere in Fountain Valley. Levels as high as twenty-five milligauss were measured at a public playground directly under the two-hundred-and-twenty-thousand-volt lines in the right-of-way; levels of fifteen milligauss were measured near houses built alongside the high-current feeder line on the north side of the power-line corridor; and similar levels were found in a number of neighborhoods traversed by other high-current distribution wires. Since no studies had been conducted of the health experience of children who attended schools near the power-line right-of-way, or who lived near the right-of-way or in the vicinity of other high-voltage or high-current lines, it was not possible to assess the extent of the hazard. However, a preliminary survey gave some cause for concern. An eleven-year-old girl at the Fulton School, who had previously attended Courreges, and lived near a high-voltage line, had been afflicted with ovarian cancer, and had undergone several operations — including a hysterectomy, and chemotherapy. A two-year-old girl who lived directly opposite the Courreges School, in a house next to the high-current feeder line in the right-of-way, had died of leukemia. And a five-year-old boy living in a house close to a high-current feeder line on Magnolia Street — a main thoroughfare — had suffered cancer of the eye, and lost the eye.

Farther south, in La Jolla — a wealthy community of thirty thousand inhabitants, which is on the Pacific Coast, about fifteen miles north of San Diego — concern expressed by the parents of an eight-year-old girl, who had developed precancerous lesions on her head, resulted in the San Diego Gas & Electric Company's measuring magnetic-field strengths at the four-hundred-and-seventy-pupil Bird Rock Elementary School, where the child was a student. The school is within a few feet of a pair of high-current distribution wires that run along La Jolla Hermosa Drive. In spite of the fact that a level of more than five milligauss was found in the school auditorium, and a level of nearly four milligauss at the kindergarten jungle gym — both of which are situated on the side of the school nearest the power lines — company officials assured the parents of children attending the school that the readings were "quite low," posed no hazard, and could be found throughout the community. In April of 1990, Stellan Knöös, a physicist and engineer, two of whose children attended the Bird Rock School, measured magnetic fields there, using a state-of-the-art gaussmeter manufactured in Sweden. Knöös, who was accompanied by Douglas Adams, the safety coordinator for the San Diego Unified School District, not only corroborated the measurements that had been taken by San Diego Gas & Electric but also found magnetic-field levels of eight milligauss at the typical eye position of children using color-display monitors in the school's crowded computer room. (Measurements of some color-display monitors have shown that many of them routinely emit levels of over four milligauss at a distance of twelve inches from their screens, and up to fifteen milligauss at the same distance from their sides.)

Early in May, Knöös, again accompanied by Adams, measured magnetic-field levels of between six and almost eighteen milligauss in several classrooms on one side of the Brooklyn Elementary School, which is in the Golden Hill section of San Diego, just a few feet from a high-current feeder line that runs along Fern Street. In a report to the school district on May 17th, Knöös recommended that classrooms with magnetic-field levels between two and four milligauss be marked for limited use only, that classrooms with levels above six milligauss be closed immediately, and that a thorough analysis be made of the health records of the children who attended the school.

Instead of taking any action to reduce the exposure of children at the Brooklyn Elementary to magnetic-field levels that were up to nine times as strong as those that had been associated with the development of childhood cancer, the San Diego Unified School District made common cause with the San Diego Gas & Electric Company, and embarked upon a two-year voyage of delay, denial, and obfuscation. On June 6th, John Dawsey, the Environmental Health Administrator for San Diego Gas & Electric, told members of the school district's newly formed Electric and Magnetic Fields Committee that the studies of the power-line health hazard that had been conducted so far were based chiefly on hypothesis and not on fact. In a memorandum circulated to school authorities on the same day, Adams claimed that because no safe levels for magnetic-field exposure had been established by the scientific community, the readings that Knöös had taken at the two schools "are not definitive and serve for information purposes only." Adams also said that the school district had a responsibility "to prevent hysteria and overreaction to issues or concerns which are still unclear or unproven."

At a meeting of the committee held on January 31, 1991, Adams again declared that not enough scientific information had been compiled to determine a safe level of exposure to power-line magnetic fields. Apparently, he believed that until such a level was established the children attending Brooklyn Elementary could be exposed with impunity to any level. Dawsey, who was attending committee meetings on a regular basis, advised committee members that Florida allowed magnetic fields of up to two hundred and fifty milligauss to exist at the edge of high-voltage power-line rights of way.

At a committee meeting held on January 17, 1992, Adams said that the scientific community was no closer to defining a safe level of exposure than it had been a year earlier. He went on to say that burying the high-current feeder line on Fern Street would not only be enormously expensive, but also would not reduce magnetic-field emissions. (In fact, as could easily have been ascertained from a June 1989 report issued by the Empire State Electric Energy Research Corporation, of New York, and as officials of San Diego Gas & Electric and those of every other utility in the nation well know, the magnetic fields given off by high-current and high-voltage power lines can be drastically curtailed if the lines are buried correctly.) At the same meeting, Jane Senour, the principal of Brooklyn Elementary, said that some of the students on the Fern Street side of the school were kindergarteners, and that she would like to move them to different classrooms but had not done so because their regular classrooms had been designed with special features for kindergarten use, and the other classrooms did not have these features. According to Harold F. Tyvoll, an attorney and resident of the Golden Hill section, who attended the January 17th meeting, Senour went on to say that she had given the kindergarten teachers the option of rotating on an annual basis, so that they could work in classrooms that were not near the power lines. She did not say why she thought it might be safe for kindergarteners to undergo daily exposure over an entire school year to magnetic fields of up to eighteen milligauss. Subsequently, Senour told Ron Ottinger, a member of the San Diego School board, that she had offered only those teachers working in the classroom with the eighteen-milligauss reading the option of rotating. Senour also told Ottinger that a teacher with medical problems had requested a transfer.

On May 6, 1992, Tyvoll wrote to Susan A. Davis, the vice-president of the San Diego Board of Education, reminding her that two years after Knöös had measured strong magnetic fields in Brooklyn Elementary classrooms, "Nothing has changed for the students, who continue to play at recess on the portion of the playground closest to the powerlines, and continue to occupy the classrooms on the east side of the school, where the radiation levels are the highest." Tyvoll pointed out that the San Diego Unified School District could avail itself of "well-recognized legal remedies" to force San Diego Gas & Electric to remove the high-current feeder line on Fern Street. He also told Davis that because the interests of the utility were clearly adverse to those of the school district, it hardly seemed proper for Adams and his colleagues on the school district's electromagnetic-field committee to be soliciting advice about the power-line health hazard from Dawsey and San Diego Gas & Electric.

Davis replied to Tyvoll in a letter dated May 29th, which contained a telling statement about the symbiotic relationship that had come to exist between the school district and the utility. "We see no conflict of interest by inviting San Diego Gas & Electric representatives to attend the EMF committee meetings since any conclusive findings made by regulatory agencies may affect how the company provides service to the district," Davis wrote.

An equally serious situation existed on the other side of the continent. During the late spring of 1990, strong magnetic fields were measured at several schools that were situated close to high-voltage transmission lines in Essex County, New Jersey, just west of New York City. Levels of between eight and eighteen milligauss were recorded on the west side of the Gould Elementary School in North Caldwell, which is about a hundred feet from a right-of-way containing a two-hundred-and-thirty-thousand-volt line and a one-hundred-and-fifteen-thousand-volt line, and levels of between three and eight milligauss were measured in the school's playground. A level of nearly thirty milligauss was measured on the north side of the Essex County Vocational

Technical High School, in West Caldwell, which is about
eighty feet from a two-hundred-and-thirty-thousand-volt
line. A level of ten milligauss was measured at the west side
of The Children's Institute, a school in Livingston for emo-
tionally disturbed and autistic children, and a level of
twenty milligauss was measured on the school's east side.
The school is sandwiched between two rights-of-way, each
of which contains a two-hundred-and-thirty-thousand-volt
transmission line. A level of twenty-five milligauss was
measured at the north side of the Burnet Hill Elementary
School in Livingston, which is about thirty feet from a right-
of-way containing a pair of two-hundred-and-thirty-thousand-
volt lines, and a level of ten milligauss was measured in a
hallway on that side of the school.

When the principal at Burnet Hill was told of the readings at his school, he issued an order forbidding the children in his charge to play in a playground located next to
the power lines. He also requested that the Public Service
Gas & Electric Company, of Newark — the owner and op-
erator of the transmission lines — take magnetic-field measurements at the school. An engineer from the utility
performed these measurements on September 11th, and on
September 17th Public Service Gas & Electric sent the principal a report showing that magnetic-field levels of nearly
thirty-three milligauss had been recorded near a basketball
hoop in the playground, and that levels of between two and
nearly eight milligauss could be found in classrooms and
other areas of the school that were used by students. A
cover letter informed the principal that "all of the measurements within the school are typical of those found in
some private homes." Shortly thereafter, articles about the
magnetic-field hazard in New Jersey schools appeared in
Family Circle magazine and the *Newark Star-Ledger*. In

spite of concern expressed by parents of children attending these schools, officials of the New Jersey State Health Department showed no interest in investigating the situation.

By the spring of 1990, however, there was a substantial body of medical and scientific evidence to show that exposure to the magnetic fields given off by power lines presented an extraordinary health hazard. The danger posed by substations and nearby high-current feeder lines appeared to be particularly acute. Nancy Wertheimer, in her investigation of the association between alternating-current magnetic fields and childhood cancer deaths had discovered that of the six children in her study population who had lived within five hundred feet of a power substation and within a hundred and thirty feet of a high-current feeder line coming from the substation, all were cancer victims — four of leukemia, one of a nervous-system tumor, and one of a sarcoma. Since her study population consisted of three hundred and forty-four children who had died of cancer and a matched control group of three hundred and forty-four living children, Wertheimer believed this to be a highly unusual finding. "Although these numbers are small, they are striking," she wrote in the March, 1979, issue of the *American Journal of Epidemiology*, and pointed out that "each cancer case had lived at the substation address within three years or less of his illness."

Wertheimer's finding seemed all the more striking in view of the extraordinary incidence of cancer among the residents of Meadow Street, the residents of Dukeville Road, and the children attending the Montecito Union School. The cancer victims on Meadow Street and in the Montecito school were spending a good part of each day within a hunded feet or so of a power substation and within fifty feet or so of high-current wires leading from the sub-station, while most of the cancer victims in the Trading Ford-Dukeville area had either worked at the Buck Steam Plant or lived in the company village, both of which were adjacent to a large substation and to more than half a dozen high-voltage transmission lines that were surely giving off very strong electric and magnetic fields. In all three situations, people were afflicted with brain cancer and brain tumors or with leukemia and lymphoma at rates far greater than the expected incidence of these diseases in the general population.

Since many, and perhaps most, of the residents of Meadow Street and Dukeville were undoubtedly subjected daily to magnetic-field levels approximating those to which electricians, power-station operators, and power and telephone linemen are exposed, and since children attending the Montecito Union School were similarly exposed to magnetic-field levels approximately half as strong as those associated with a seven-fold increase of leukemia among telephone-company cable splicers, some observers thought it appropriate to point out that during the previous ten years nearly two dozen epidemiological studies had been conducted and published in the peer-reviewed medical literature here and elsewhere in the world, showing that electricians, power-station operators, power and telephone linemen, and other workers whose occupations exposed them to electromagnetic fields were developing and dying of leukemia, lymphoma, brain cancer, and brain tumors at rates significantly higher than those observed in unexposed workers. The first of these studies was conducted by Dr. Samuel Milham, Jr., a physician and epidemiologist with the Washington State Department of Social and Health Services, in Olympia. Milham examined the data for four hundred and thirty-eight thousand deaths occurring among working-men in Washington between 1950 and 1979, and he noticed that among men whose occupations required them to work in electric or magnetic fields the ratio of deaths caused by leukemia was higher than the proportionate mortality ratio in ten out of eleven occupations he investigated. Milham's findings, which were published as a letter in the *New England Journal of Medicine* in July of 1982, have since been supported by similar findings among electrical workers in Los Angeles, New Zealand, Canada, and south-eastern England.

Even more striking had been the growing evidence of a link between such exposure and brain tumors. This link was examined in some detail by Louis Slesin, the editor and publisher of *Microwave News*, which is a newsletter that reports on nonionizing radiation and is published six times a year in New York. In the March-April, 1990, issue of the newsletter, Slesin pointed out that, according to a study in the medical journal *The Lancet*, brain-cancer deaths had nearly tripled among older white men and women in the United States between 1968 and 1983, and he went on to list no fewer than twelve studies conducted, published, or reanalyzed between 1985 and 1989 that showed significantly increased rates of brain tumors among people exposed to electric and magnetic fields at home or at work. Among the studies were an investigation revealing that a higher than expected number of white male residents of Maryland who were employed as electricians, electrical and electronics engineers, and utility servicemen had died of brain tumors between 1969 and 1982; an analysis by Dr. Milham of data from his earlier survey showing that electricians in Washington State experienced a fifty-five per cent greater risk of dying of brain tumors than other workers;

and a 1988 study of people who had died of brain cancer in East Texas between 1969 and 1978 which showed that the risk for electric-utility workers was thirteen times that of workers who were not exposed to electromagnetic fields. Also included was the finding by Genevieve Matanoski and her associates at Johns Hopkins that telephone-company cable splicers are afflicted with brain tumors at almost twice the rate of other workers; a 1989 study by Susan Preston-Martin and some colleagues at the University of Southern California School of Medicine, in Los Angeles, showing that men with high exposure to electric and magnetic fields were more likely than other workers to develop brain cancers, such as gliomas, particularly astrocytomas; an analysis of data from the earlier childhood-cancer study by David Savitz and his colleagues, which found that children living near high-current power lines are almost twice as likely to develop brain tumors as children living near low-current wires; and another finding by Savitz and an associate that electrical and electronic technicians develop brain tumors at three times the rate of workers who are not exposed to electromagnetic fields, while electric-power repairmen and installation workers develop such tumors at more than twice that rate. Particularly disturbing were the results of a study published in 1989 in the *International Journal of Epidemiology* by Christine Cole Johnson, the head of epidemiology at the Division of Biostatistics and Research Epidemiology of the Henry Ford Hospital, in Detroit, and by Margaret R. Spitz, an epidemiologist in the Department of Cancer Prevention of the University of Texas M. D. Anderson Cancer Center in Houston. They found that children whose fathers were electricians ran three and a half times the risk of developing tumors of the central nervous system that other children ran; indeed, tumors of the brain stem had been diagnosed in no fewer than four of seven children of electricians in their study population.

As for the ways that exposure to electric and magnetic fields might adversely affect the central nervous system and the brain, and either cause or promote cancer, there was a wealth of information on that subject, starting with a 1965 study conducted by Dr. Robert O. Becker, an orthopedic surgeon at the Veterans Administration Hospital in Syracuse, New York, and Howard Friedman, a psychologist at the hospital, which showed that exposure to strong pulsed magnetic fields considerably slowed the reaction times of human volunteers. In 1964, a Soviet investigator named Yuri Alexandrei Kholodov had reported that exposure to strong magnetic fields produced areas of cell death in the brains of rabbits. Shortly thereafter, other Soviet scientists reported that workers in high-voltage switchyards were experiencing fatigue, drowsiness, headaches, and other symptoms of central-nervous-system stress. During the nineteen-seventies, Dr. W. Ross Adey, a neurological scientist, who was then the director of the Space Biology Laboratory at the Brain Research Institute of the University of California at Los Angeles, conducted a series of experiments demonstrating that weak electromagnetic fields oscillating at extra-low frequencies could significantly alter the chemistry of the cerebral tissue of chickens and of the brains of living cats.

During the early nineteen-eighties, Adey, who had become associate chief of staff for research and development at the Jerry L. Pettis Memorial Veterans' Hospital, in Loma Linda, California, conducted experiments with some colleagues which demonstrated that electromagnetic fields pulsed at the electrical-distribution-system frequency of sixty hertz could inhibit the ability of cultured T-lymphocyte cells from mice to kill cultured cancer cells — a result suggesting that these fields might be acting as cancer promoters by suppressing the immune system. Subsequently, an associate of Adey's used simulated sixty-hertz high-voltage power-line fields to produce the same result. Also worrisome were the results of a 1986 experiment in which Adey and his colleagues found that a one-hour exposure to a sixty-hertz electric field of between one-tenth of a millivolt and ten millivolts per centimetre produced a fivefold increase in the activity of an enzyme known to be associated with the development of tumors.

Following the publication of these results in the journal *Carcinogenesis*, in October of 1987, Adey said that he and his colleagues could now theorize that "exposure to low-energy fields, such as those emanating from power lines, may provide a tumor-promoting stimulus." This was an ominous hypothesis, because an alternating-current electric field of one-tenth of a millivolt is present at all times in the tissue of a human being who is standing beneath a typical overhead high-voltage transmission line. It became additionally ominous in the light of a study conducted in 1987 by Reba Goodman, a geneticist and cell biologist at Columbia University's Medical Center, in New York, and Ann S. Henderson, a molecular biologist at Hunter College, who have demonstrated that weak, pulsed sixty-hertz electric and magnetic fields similar to those given off by power lines may trigger the development of cancer by altering RNA transcription — the way genetic instructions are carried out in organisms — to increase the production of proteins frequently found in tumor cells.

All this epidemiological and experimental evidence of the cancer-producing potential of power-line electric and magnetic fields was frighteningly relevant to the predicament of thousands upon thousands of people who, like the residents of Meadow Street, were living so close to substations and high-current wires that they were continuously exposed within their own homes to electromagnetic fields of occupational levels. In addition, there was epidemiological and experimental evidence from Sweden and the Soviet Union to suggest that electromagnetic fields could cause fertility problems among men working in high-voltage substations and switchyards, and an increased rate of birth defects among their children — findings that certainly appeared to be relevant to the experience of many of the residents of 48 Meadow over the past thirty-five years. For example, in 1983, Professor S. Nordström, of the Department of Public Health and Environmental Studies of the University of Umeå, in Sweden, and some colleagues reported in the medical journal *Bioelectromagnetics* that they had found a significant increase in congenital anomalies among the children of men who worked in high-voltage switchyards and were thus being constantly "electrically charged and then discharged," causing "cell disturbances, including chromosomal aberrations." Laboratory studies of lymphocytes from a sample of high-voltage substation workers in Sweden indicated that chromosome breaks and aberrant cells are significantly more common among these workers than among a control group of unexposed men. Other studies suggested that such effects might be caused by exposure to the electrical pulses (spark discharges) that are given off by transformers and other equipment in substations, rather than to the sixty-hertz electromagnetic fields. In this connection, a study conducted by researchers at McGill University's School of Occupational Health, in Montreal, showed that certain electric-utility workers (and so, quite probably, the residents of Meadow Street) have

been exposed to ten times the ordinary background level
of sixty-hertz electric and magnetic fields, and up to a
hundred and seventy times the background level of the
high-frequency transient electromagnetic fields that are
common in the vicinity of substations and other facilities
where switching operations are conducted.

The response of the electric-utility industry to this growing
body of evidence scarcely seemed designed to enlighten
the public about the power-line hazard. In 1989, the Edison
Electric Institute, of Washington, D.C., an association of electric companies, published a brochure stating that measurements of behavior and brain function in animals exposed to
power-line electromagnetic fields showed "some small effects that may be the result of body rhythm changes," but
that "the connection between these laboratory results and
human health is not known." The credibility of this assertion was soon placed in doubt by news that a team of experimental psychologists had found that men subjected to
strong electromagnetic fields exhibited motor responses
ten per cent slower than those of their unexposed counterparts, and slower heartbeat rates and brain-wave patterns as
well. The Virginia Power Company and the North Carolina
Power Company published a newsletter offering the assurance that "there is always some electrical current in everyone's body and it is necessary for life," and that "electric
and magnetic fields induce small additional amounts." In a
pamphlet mailed to consumers in 1989, the Commonwealth Electric Company, of Massachusetts, acknowledged
the existence of an epidemiological study showing that
childhood-cancer victims are "more likely" (the study's actual finding was twice as likely) to have high-current power
lines outside their homes than are children without cancer,

but it added, "Just what the results of this study mean is not at all clear." The authors of a bulletin issued by the Southern California Edison Company — no doubt written before the cancer cluster at the Montecito Union School came to light — posed the question of whether magnetic fields given off by power lines are harmful to human health, and supplied an answer that contained an interesting euphemism for cancer. "There is no proven health hazard from electric and magnetic fields from electric utility facilities," they declared. "However, some studies *suggest* that there may be an association between either magnetic or electric fields and certain health risks."

The response of the federal government to growing evidence that the magnetic fields from power lines posed a public-health hazard had been cautious at best and irresponsible at worst. Thanks to budget cuts imposed by the Reagan Administration, a promising program of investigation into the biological effects of extra-low-frequency (ELF) electromagnetic fields, which had been under way for ten years at the Environmental Protection Agency's Health Effects Research Laboratory, at Research Triangle Park, in North Carolina, was shut down in 1986. Since that time, almost no funding had been made available by the E.P.A. for research in this area, and federal-government involvement had been limited to a modest research program financed by the Department of Energy, which (as could be seen from its gross mismanagement of the nuclear plants under its jurisdiction) had shown little interest in protecting the public health.

As for the state health agencies, their reaction to the rapidly emerging power-line hazard had been typified by the desultory efforts of Dr. Morris and his colleagues in North Carolina with regard to the brain-tumor situation at Trading

Ford and Dukeville, and by the similarly inadequate response of Dr. Neutra and his associates in California to the cancer cluster that had developed among children attending the Montecito Union School. Still another example would be provided by the public health authorities in Connecticut.

CHAPTER SIX

Two One-in-Fifty-Thousand Chances

A T A PUBLIC MEETING held in the Guilford Public Library on August 20, 1990, David R. Brown, chief of the Connecticut Department of Health Services' Division of Environmental Epidemiology and Occupational Health, and Sandy Geschwind, an epidemiologist with the division, appeared on the same program with officials and paid consultants of Northeast Utilities, of Hartford — parent of the Connecticut Light & Power — and told a hundred or so Guilford residents that there was no cancer cluster on Meadow Street. To support their contention, they distributed a document entitled "Guilford Cancer Cluster Preliminary Investigation," claiming that "there was not a cluster of the same kind of tumors on Meadow Street," and that from 1968 through 1988 "Guilford as a whole did not experience a higher than expected number of brain cancer or meningioma cases." The document stated, further, that "mapping of these brain tumor and meningioma cases showed that they did not cluster in a particular area but were scattered throughout the town."

At that meeting, Geschwind gave a presentation in which

she said that one of the brain cancers on Meadow Street was not a primary tumor but an esophageal cancer that had metastasized. She also said that eye melanoma was a type of cancer that had never been associated with exposure to electromagnetic fields, and she assured her listeners that meningioma had never been associated with exposure to such fields. Toward the end of her presentation, Geschwind displayed a map showing the location of ten meningiomas and nineteen other brain and central-nervous-system tumors listed by the Connecticut Tumor Registry as having occurred in Guilford between 1968 and 1988, and told the hundred or so members of her audience — they included a dozen newspaper and television reporters — that the map proved that there was "absolutely no clustering" in Guilford and that the state investigation showed "no cancer cluster on Meadow Street."

However, the fact that Guilford as a whole — the town had a population of twenty thousand five hundred, living in seventy-three hundred dwellings — did not experience a higher than expected number of meningiomas and other brain and nervous-system tumors during those twenty-one years did not address the situation of Meadow Street. Second, while there was no reason to doubt Geschwind's assertion that one of the two brain cancers among Meadow Street residents was not a primary tumor, eye melanoma — the one in question was a malignant tumor involving the optic nerve, an extension of the brain — had been found to be "notably high for electrical and electronics workers," who are known to be exposed to strong magnetic fields. The finding appeared in a highly regarded study entitled "Epidemiology of Eye Cancer in Adults in England and Wales, 1962–1977," which was conducted by Dr. A. J. Swerdlow, a physician at the Department of Community

Medicine of the University of Glasgow, in Scotland. Swerdlow reported his findings in 1983, in Volume 118, No. 2,
of the *American Journal of Epidemiology*. Moreover, melanoma of the skin was one of three types of cancer listed
by scientists of the Environmental Protection Agency in a
draft report, "An Evaluation of the Potential Carcinogenicity
of Electromagnetic Fields," as being prevalent among workers in electrical and electronic occupations, and thus associated with exposure to magnetic fields.

The conclusion of Brown and Geschwind that there was
no cancer cluster among people who had lived on Meadow
Street seemed disingenuous, to say the least. As Geschwind
noted, the Connecticut Tumor Registry recorded ten cases
of meningioma and nineteen other primary tumors of the
brain and central nervous system among Guilford residents
between 1968 and 1988 — a span in which the average population of the town was seventeen thousand five hundred.
Thus the meningioma rate in Guilford was consistent with
the Connecticut statewide incidence, of 2.6 cases per
hundred thousand people per year, and the incidence of
other brain and central-nervous-system tumors in Guilford
was also close to the number that would normally be expected. The fact that three of the twenty-nine primary brain
and central-nervous-system tumors that occurred in Guilford during those twenty-one years developed among a
handful of people who lived in four of five adjacent houses
on Meadow Street that are situated near a substation and
very close to a pair of high-current distribution feeder
lines, together with the fact that a malignant eye tumor, involving a tract of brain tissue, occurred in a woman who
had lived in a sixth adjacent dwelling, next to a third feeder
line, surely suggested that there was a cancer cluster of
some significance on Meadow Street.

Finally, and somewhat ironically, further evidence of cancer clustering associated with exposure to power-line magnetic fields could be found in the very map that Geschwind displayed in an effort to persuade the people of Guilford that no cancer cluster existed there. Among those listening to her presentation was Robert Hemstock, the Guilford resident who had first sounded the alarm about a cluster on Meadow Street. When Geschwind held up the map, Hemstock noticed that three of the twenty-nine cancers on it appeared to have occurred along the route of a feeder line that carried high current from the Meadow Street substation to other towns during the nineteen-sixties, seventies, and early eighties, when the substation was being operated by its owner, the Connecticut Light & Power Company, as a bulk-supply station for large-load areas in Madison and Clinton — neighboring towns with a total population of about twenty thousand during that period. He also noticed that an unusually large proportion of the other brain tumors on the map appeared to have occurred among people living along the routes of other primary distribution lines emanating from the substation.

After the meeting, Hemstock shared his observation with Don Michak, a reporter for the *Manchester Enfield Journal Inquirer,* who on August 23rd asked the Department of Health Services for a copy of the map. As it happened, Brown had displayed the map the day before at a Rotary Club meeting in Guilford, and told the Rotarians that he saw no need for the department to make any further inquiry into the incidence of cancer on Meadow Street. However, Health Services officials refused to release the map to Michak, on the ground that to do so might violate the confidentiality of cancer victims by revealing their addresses. The *Journal Inquirer* reported this development in an article by Michak on September 6th, and on September 10th it published an editorial pointing out that if the with-held map showed that the distribution of cancer cases in Guilford corresponded to the Meadow Street substation and to a power line running north from it "the public's concern might be overwhelming not only in Guilford but throughout Connecticut and even nationally." The editorial went on to question Health Services' rationale for secrecy, declaring that the map "is just a matter of dots superimposed on a map of Guilford; it apparently doesn't include names and addresses," and that "anyone seeking to use the map to find people who have or had cancer would have to knock on doors in the area of the dots on the map and ask such people to identify themselves." After observing that "the health department undermined its own rationale by displaying the map at the public hearing in Guilford in the first place," the editorial concluded by stating that if the department failed to make the map available "the public will have to assume that the department wants to protect something else more than it wants to protect public health."

In September, a reporter for the *New Haven Register* obtained a copy of the map from an assistant to the Guilford health officer. (The assistant later said that she had given it out by mistake.) The *Register* reporter also went to the Connecticut Light & Power Company's office in Madison and obtained a company map of the routes of existing high-current and high-voltage distribution lines in Guilford. On October 3rd, the *Register* published its own map — one combining the locations of the brain tumors and other central-nervous-system tumors with the routes of Connect-icut Light & Power's distribution lines. It clearly showed that Hemstock's observation was correct — that an inordinately high number of the meningiomas and other brain and central-nervous-system tumors that had occurred in Guilford over the twenty-one-year period between 1968 and 1988 had developed in people living close to primary distribution wires.

This correlation notwithstanding, Brown and Geschwind denied that the map furnished any evidence of a link between the occurrence of such tumors and proximity to power lines in Guilford. "You can't use the map to show that kind of association," Geschwind told the *Register*. She added that such tumors could be found on streets near main distribution power lines because those streets were densely populated, and heavily populated areas would have proportionally higher cancer rates.

Hemstock soon proved her wrong. After obtaining the addresses of the twenty-nine brain and other central-nervous-system tumor victims in Guilford, he and some colleagues followed the routes of the feeder lines and primary distribution wires leading from the Meadow Street substation, and they found not only that there was a strong correlation between the occurrence of these tumors and living close to high-current or high-voltage wires but also that most of the tumors had not occurred in areas of notably dense population. The feeder that carried high current from the substation to Madison and Clinton was abandoned a few years ago; it ran across Meadow Street from the substation and proceeded east for about a mile and a half, to a point near the junction of Stone House Lane, South Union Street, and Sawpit Road. (Up to that point, the poles and the wires of the line remain in place, but they have been removed from the rest of the route — across an uninhabited salt marsh and the East River, which is the eastern boundary of Guilford, to a substation on Garnet Park Road, in Madison.) This feeder line ran for a mile and a half through Guilford, and it passed close — within a hundred and fifty feet or so — to only twelve houses. One of the ten meningiomas and two of the nineteen other brain and central-nervous-system tumors listed by the Tumor Registry as having occurred in Guilford between 1968 and 1988 afflicted people living in three of those twelve dwellings. All three are situated within about forty feet of the high-current wires. Moreover, Judith Beauvais, the former Meadow Street resident who developed eye cancer at the age of forty-four and died of it, lived for fourteen years in one of the twelve houses close to the abandoned feeder line. It is at 56 Meadow, and is situated only about thirty feet from the wires.

Four other feeder lines carry or previously carried high current from the Meadow Street substation to various sections of Guilford. Among them are a pair of thirteen-thousand-eight-hundred-volt lines that run northwest across marshland at the rear of the substation to Water Street near the West River. (Earlier in 1990, spreaders had been placed on these and the other thirteen-thousand-eight-hundred-volt feeder lines, pulling the wires into a close configuration and reducing the magnetic fields that surround them.) One of these lines turns west on Water Street and continues along Leetes Island Road to the Guilford-Branford line. The other runs east along Water Street to Guilford Green, in the center of town; continues east on Boston Street; and then follows the Boston Post Road to the Guilford-Madison line. The combined length of the two lines is about ten and a half miles, and over this distance their high-current wires come within a hundred and fifty feet or so of a hundred and sixty-two houses out of a total of a hundred and seventy-three houses that are visible from the road. One of the nineteen brain and central-nervous-system tumors listed by the Connecticut Tumor Registry occurred in a resident of a dwelling situated within fifty feet of the wires.

At the southeast corner of Guilford Green, a primary distribution line branches off that feeder line and runs north along State Street, north on Nut Plains Road, and northeast on North Madison Road, to the Guilford-Madison boundary. This line is about five and a half miles long and passes within a hundred and fifty feet of a hundred and seventy-two out of a total of a hundred and seventy-nine houses. Another of the nineteen tumors occurred in a resident of a dwelling close to the wires.

A third thirteen-thousand-eight-hundred-volt feeder line runs north from the substation along Meadow Street and across Water Street. It continues north along River Street and then follows the Boston Post Road in a northwesterly direction to the Guilford-North Branford line. The length of this line is almost four miles. Along its distance, it passes within a hundred and fifty feet of eighty-two houses out of a total of ninety-four houses that are visible from the roadway. None of the meningiomas or other brain and central-nervous-system tumors listed by the Tumor Registry occurred in residents of these dwellings.

The fourth feeder line — this one carrying twenty-seven thousand six hundred volts — follows the same route from the Meadow Street substation to the junction of the Boston Post Road and Long Hill Road. It then runs north along Long Hill Road to Hubbard Road, and there delivers current to a paper-coating factory that employs some two hundred workers. (The loading-dock supervisor at the factory was found to have developed brain cancer a week after Hemstock and his colleagues conducted their survey; his office was next to a high-voltage transformer where

Hemstock had measured a magnetic field of thirteen milligauss.) A continuation of this line carries thirteen thousand eight hundred volts north along Long Hill Road to Bullard Drive. The total length of the line from the Meadow Street substation is slightly more than five and a half miles, and over that distance it passes within a hundred and fifty feet of a hundred and thirty-four out of a total of a hundred and sixty dwellings. Including the tumors on Meadow Street, five of the ten meningiomas and one of the nineteen other brain and central-nervous-system tumors listed by the Registry have afflicted people living in houses close to this line.

In their analysis of the association between the development of such tumors and the proximity of residences to main distribution lines, Hemstock and his associates also considered branch lines carrying high current from some of the feeders that originate at the Meadow Street substation. One such branch leaves the line on Long Hill Road, runs northeast on Flat Meadow Road, and then follows Durham Road (Route 77) to the north for a mile and to the south for about a mile and a half. The length of this line is approximately three miles, and it passes within a hundred and fifty feet of fifty-one out of a total of fifty-four houses. Two of the nineteen brain and other central-nervous-system tumors developed in residents of houses close to this line. Another branch runs east from River Street along the Boston Post Road for about two miles. It passes close to many places of business in the center of Guilford but close to fewer than a dozen residences, and no meningiomas or other brain tumors were reported by the Registry as having occurred among people living along its route.

In addition to the five feeder lines carrying high current from the Meadow Street substation, a main distribution line originating at a substation in North Madison runs west into Guilford along Route 80 and then south on Flat Iron Road to Edgehill Road. The total length of this line in Guilford is about five miles, and over that distance it passes within a hundred and fifty feet of seventy-three out of a total of ninety-eight dwellings. Two of the nineteen brain and other central-nervous-system tumors occurred in residents of houses close to the wires.

Finally, Hemstock and his colleagues traced the routes of the four high-voltage transmission lines that run through Guilford. Three of them come from a substation in Branford and run east through Guilford for about three miles. One of these lines was abandoned a few years ago, but the two others still supply voltage to the Meadow Street substation. These lines pass close to fewer than a dozen houses in Guilford and no cases of brain tumor among their residents were reported by the Registry. A fourth transmission line, carrying a hundred and fifteen thousand volts, comes from the same substation, crosses Guilford from west to east, and supplies voltage to substations in Madison and Old Saybrook. This line gives off strong magnetic fields, but since it runs cross-country for most of its five-mile length in Guilford there are only about a dozen houses situated close to it. Nevertheless, a resident of a dwelling about seventy-five feet from the line has developed an astrocytoma — a malignant brain tumor — and the resident of another house near the line developed a meningioma.

All told, seven of the ten meningiomas and ten of the nineteen other brain and central-nervous-system tumors — that is, seventeen of the total of twenty-nine, nearly sixty per cent — afflicted people living near high-current or high-voltage power lines in Guilford. The total combined length of the lines is about forty-five miles, and along this distance some seven hundred and twenty-two out of a total of eight hundred and six houses are situated within a hundred and fifty feet of the wires. It seemed obvious to Hemstock and his colleagues that in a town of seventy-three hundred dwellings the occurrence of this proportion of meningiomas and other brain and central-nervous-system tumors in residents of just over eight hundred dwellings strung out along some forty-five miles of roadway could not be ascribed to heavy population — as the Connecticut Department of Health Services had done. It also seemed obvious that people living in houses close to high-current wires and high-voltage transmission lines in Guilford were especially susceptible to developing meningiomas and other brain tumors. Particularly disturbing in this regard was Hemstock's discovery that in March of 1989 — too late to be counted among the twenty-nine tumors listed by the Registry on the map that the Connecticut Department of Health Services displayed to reassure the townspeople of Guilford — a seventeen-year-old girl living in a house close to one of the high-current feeder lines was found to be suffering from an astrocytoma, the same type of malignant brain tumor that had been diagnosed two months earlier in seventeen-year-old Melissa Bullock, who lived near the very same line on Meadow Street. Considering the fact that astrocytoma is expected to develop in about one in every fifty thousand seventeen-year-old women each year, the odds that such a calamity could have occurred by chance almost simultaneously to these two young women seemed vanishingly small.

In November of 1990, shortly after the results of the survey Hemstock and his colleagues had conducted became known, there were calls for the Connecticut Department of Health Services to conduct a thorough study of the apparent strong association between the occurrence of astrocytomas, meningiomas, and other brain and central-nervous-system tumors, on the one hand, and on the other, chronic exposure to the magnetic fields given off by high-current and high-voltage power lines in Guilford. Moreover, since Connecticut was one of the few states that had collected data on the occurrence of such tumors over a significant period, it was suggested that the Department had a unique opportunity to perform an important service for public health nationwide by conducting a detailed investigation of the seventeen hundred and three meningiomas and the four thousand one hundred and two other brain and central-nervous-system tumors that had been diagnosed among Connecticut residents over the twenty-one years between 1968 and 1988, in order to determine whether, as was clearly the case in Guilford, a disproportionately high percentage of them had developed in people living close to wires giving off strong magnetic fields. However, Department officials rejected these proposals out of hand.

PART TWO

THE CANCER
AT SLATER SCHOOL

CHAPTER SEVEN

The Kiss of Death

A FEW WEEKS LATER, on a Friday afternoon in mid-December, half a dozen women who taught at the Louis N. Slater Elementary School in Fresno, California — a city of nearly four hundred thousand inhabitants that is situated in the San Joaquin Valley, in the central part of the state — were interviewed in the teacher's lounge of the school by Amy Alexander, a staff writer for the *Fresno Bee,* who wanted to know if they were concerned about the presence of a pair of high-voltage transmission lines that ran past the school on Emerson Avenue. That morning, the *Bee* had published an Associated Press story about an attempt by the Bush Administration to delay the release of a report compiled by the federal Environmental Protection Agency, which linked residential and occupational exposure to the alternating-current magnetic fields given off by power lines with the development of cancer in children and adults. Earlier in the week, a parent whose child attended the Tobey B. Lawless Elementary School, about a mile to the northwest of Slater, had asked the principal there to get in touch with the Pacific Gas & Electric

Company, of San Francisco — the utility that serves Fresno and is the largest in the nation — about the potential hazard of some high-voltage transmission lines that run along Corona Avenue, within three hundred feet of the school. The parent also mentioned her concern to an acquaintance on the *Bee*'s staff, who passed it along to Alexander. After making a few telephone calls, Alexander learned that the Slater School sits only a hundred feet or so from the same high-voltage power lines that run past Lawless. It was then that she arranged to meet with teachers at Slater.

Up to the time of Alexander's visit, none of the forty-three teachers at the Slater School had apparently ever entertained any doubts about working close to power lines. When Alexander told the women she met with in the teacher's lounge about the E.P.A. report, however, they were quick to inform her that an unusually large number of teachers and teacher's aides at Slater had developed cancer in recent years. Two days later, an article by Alexander about the potential hazard at Slater appeared on the front page of the Sunday edition of the *Bee,* under a headline that read, "POWER LINES WORRY SCHOOL." It was accompanied by a photograph that showed a fifty-two-year-old teacher named Sandy Craft, who had been found to have cancer of the uterus, four years earlier, leading a group of children along the sidewalk on Emerson Avenue, just across the street from an eighty-four-foot-high steel-lattice transmission tower. In her article Alexander said that the transmission lines on Emerson Avenue supplied power for more than forty thousand Fresno homes, and that transmission lines had been there since the 1920s. She also reported that since 1987 the California Department of Education had required that new schools be placed at least one hundred and fifty feet from such lines.

Earlier in the article, Alexander quoted an E.P.A. official in San Francisco, who said that President Bush's national science adviser had expressed reservations about the Agency's report concerning the potential carcinogenicity of power-line emissions because he feared it could cause "widespread panic." She then quoted the Slater School principal, George Marsh, who cautioned that "we need to allay people's fears rather than have them become hysterical." After reporting that P. G. & E. representatives would be taking measurements at the Slater and Lawless schools on the following day, Alexander described a spokesman for the California Department of Health Services as saying that he was unfamiliar with concerns about health risks from power lines near schools in the Fresno area, and the Fresno Unified School District's director of health services as saying that she was unaware of any unusual incidence of cancer at any of the district's schools. But Loretta Hutton, one of the women Alexander interviewed in the teacher's lounge, declared that the high incidence of cancer at Slater had not gone unremarked by some of her colleagues in the school district. "They say 'Oh, you teach at Slater? Well, that's certainly the kiss of death,' " Hutton told Alexander.

As might be expected, Alexander's article was the topic of considerable discussion among the teachers at Slater during the following week. Nothing much came of this, however, because on the following Friday the faculty and pupils went on their annual two-week Christmas vacation. Moreover, because Slater is a year-round school, a system was in effect under which teachers there are required to take two six-week vacations during the year, with the result that nearly a dozen teachers, including several whom Alexander had interviewed for her article, did not return to work until the middle of February. Among them was

Patricia Berryman, a first-grade teacher and the mother of two grown children, who had been teaching at Slater for fifteen years, and had special reason to find Alexander's piece unsettling. An attractive woman in her late forties, Berryman had been born in Fresno, gone to Fresno High School, and graduated from Fresno University, in 1964. Except for a four-year period when she lived in Los Angeles, she had taught in Fresno public schools since 1965. She had come to Slater in August of 1975, and had since been teaching reading, writing, math, science, and social studies to first-graders in Pod A — an octagonal area at the southeast corner of the school. Pod A, which faces Emerson Avenue, contains three first-grade classrooms and two kindergarten rooms. Slater has four such pods: they are situated at the four corners of the school building, and surround a large rectangular area that houses offices for the principal, the vice-principals, the administrative secretaries, and nurses, and also a kitchen, a faculty room, an all-purpose room, a patio, and several workrooms. Pod B, at the southwest corner of the building, also faces Emerson Avenue, and contains five classrooms for fifth- and sixth-graders. Pods C and D, at the northwest and northeast corners of the school, respectively, are, of course, farther away from Emerson Avenue; they contain ten classrooms for second-, third-, and fourth-graders.

"During the mid-nineteen-eighties, I began to realize that cancer and various other tumors were striking a lot of people who worked around me at Slater," Berryman said recently. "One reason for my awareness was that I was a member of the Slater Sunshine Committee — a group of three or four teachers, who visit sick staff members and send them flowers, plants, and get-well cards. In 1983, a nurse's aide, who worked in an administrative office just behind and adjacent to Pod A, developed breast cancer and had a mastectomy. The following year, a teacher's aide who had worked with me in Room 3 of Pod A for five years developed melanoma. She died two years later, after the cancer spread to her lungs. In January of 1985, a fifth-grade teacher, who had been working for two years in Pod A and for three years in Pod B, developed meningioma — a nonmalignant tumor of the brain — which impaired the sight of one eye and forced her to retire on disability. In 1986, two more of my colleagues were afflicted with cancer. One was Sandra Craft, who developed cancer of the uterus after teaching kindergarten for twelve years in the room next to mine in Pod A. The other, a fifth-grade teacher, who had worked in Pod B for eight years, developed breast cancer and had a mastectomy. In the spring of 1987 a fifth-grade teacher who had worked for fourteen years in Pod B developed ovarian cancer, and a teacher's aide in Pod B developed breast cancer. In addition to these malignancies, at least four other teachers in Pods A and B, myself among them, as well as several office and kitchen-staff workers in the adjacent administrative area, had developed nonmalignant tumors of the breast or the uterus.

"During this whole time, I knew of no one working elsewhere in the school who had been diagnosed with cancer or a tumor. Sandy Craft and I had begun to wonder why there should be so much disease among those of us who worked in or near Pods A and B. Then, in March of 1990, brain cancer was diagnosed in Katie Alexander, a close friend, who had taught first grade with me in Pod A for fifteen years. She had been having trouble with her eyes, and had gone to see an ophthalmologist. He sent her to a neurological surgeon, who, after determining her condition, operated on her brain and removed a malignant tumor the size of a baseball. During the summer and autumn of 1990, three or four of us had been visiting Katie every other week or so at her mother's house, where she was recuperating from the surgery, and in December, when I learned about the E.P.A. report and the fact that it specifically cited brain cancer as being associated with exposure to power-line magnetic fields, I began to wonder for the first time about the huge transmission lines I could see through the window of my room in Pod A, running down Emerson Avenue."

CHAPTER EIGHT

A Probable/Possible Cause of Cancer

THE E.P.A. REPORT — a three-hundred-and-sixty-seven-page document entitled "Evaluation of the Potential Carcinogenicity of Electromagnetic Fields" — had come to light in March of 1990, when someone in the Agency had sent a draft version of it to Louis Slesin, the editor and publisher of *Microwave News,* a pioneering and influential newsletter that had been covering developments in the research into the hazards of exposure to radiation from power lines and other sources since the early 1980s. Upon making inquiries, Slesin learned that on March 6th, Richard Guimond, the director of the E.P.A.'s Office of Radiation Programs, and William Farland, director of the Agency's Office of Health and Environmental Assessment — the office that had prepared the report — had briefed officials of the White House Office of Science and Technology Policy and the White House Office of Policy Development about its conclusions and recommendations. Chief among the conclusions was one specifying that power-line electromagnetic fields should be classified as a "probable" human carcinogen. Slesin also learned that after his meeting with the White House officials, Farland had ordered the paragraph containing this conclusion to be deleted from the report. The deleted paragraph read as follows:

> Concerning exposure to fields associated with 60 Hz electrical power distribution, the conclusion reached in this document is that such exposure is a "probable" carcinogen risk factor, corresponding to a "B1" degree of evidence that it is a risk factor. This conclusion is based on "limited" evidence of carcinogenicity [in] humans which is supported by laboratory research indicating that the carcinogenic response observed in humans has a biological basis, although the precise mechanisms [are] only vaguely understood.

In spite of the deletion, the summary and conclusions section of the report contained a persuasive indictment of power-line magnetic fields as a cancer-producing agent. Its authors stated that five of six case-control studies published in the peer-reviewed medical literature showed that children living in homes near power lines giving off strong magnetic fields were developing cancer more readily than children who did not live near power lines. This association was statistically significant in three of the studies; and in two studies in which magnetic-field measurements had been made children exposed to fields of between two and three milligauss or above were experiencing a significantly increased risk of developing cancer. The E.P.A. researchers declared that a "consistently repeated pattern of leukemia, nervous system cancer and lymphoma in the childhood studies" argued "in favor of a causal link" between the development of these tumors and the exposure of children to power-line magnetic fields. They went on to say that they had reviewed more than thirty reports dealing with cancer incidence or mortality among workers exposed to such fields, and found that the results of the occupational studies tended to support the results of the childhood studies, with leukemia, brain cancer, and malignant melanoma of the skin predominating among the exposed workers. "These cancer sites are found consistently across different geographic regions, age groups, industries, occupational classifications, and study designs," they wrote. "Given this diversity of studies, in addition to the likelihood that across broad job categories the exposures to various chemicals is not uniform, it is difficult to identify any single agent or group of confounding exposures that could explain the consistent finding of these same cancer sites."

After determining that magnetic fields were the most likely culprit for the excess cancer found in the epidemiological studies, the authors of the EPA report described a body of experimental research — it included studies showing that magnetic fields could impair the immune system and inhibit the synthesis of melatonin, a hormone known to be present in reduced amounts in women who develop breast cancer — and declared that there was "reason to believe that the findings of carcinogenicity in humans are biologically plausible." However, in apparent deference to the wishes of White House science and policy advisers, they went on to say that they did not consider it appropriate to classify electromagnetic fields as a cancer-producing agent, because the basic nature of interaction between the fields and the biological processes leading to cancer was not understood. The final paragraph of the revised "Summary and Conclusions" section contained this sentence: "With our current understanding, we can identify 60 Hz magnetic fields from power lines and perhaps other sources in the home as a possible, but not proven, cause of cancer in people."

As things turned out, most of the newspapers that ran

stories about the E.P.A. draft report did not describe its findings in any detail but relied upon the assessments of officials like Farland, who went to considerable lengths to play
down the power-line hazard. Typical of the coverage was a
piece that ran in the *New York Times* on May 23rd; it reported that Farland had said that the available evidence
"showed only a statistical association" between the development of cancer and exposure to electromagnetic fields.
"We are only saying that we are seeing a link that may be
significant," Farland said. He added, "The real message we
want to send is that this information should not cause
undue alarm but does suggest the need for additional research." Farland continued in this vein on July 25th, when
he told the House of Representatives Subcommittee on Natural Resources, Agriculture Research, and Environment of
the Committee on Science, Space, and Technology that the
E.P.A. was not making any recommendations to the general
public about the general hazard, and that "while the public
should be aware of this emerging scientific issue, I see no
need for alarm."

Farland's not-to-worry attitude was carried a step farther
by Paul L. Zweiacker, manager of environmental planning
for the Texas Utilities Company, of Dallas, who appeared before the subcommittee in behalf of the Edison Electric Institute (EEI), of Washington, D.C. — an association of the
nation's investor-owned utilities, which generate nearly
eighty per cent of the electricity used in the United States.
Zweiacker told the subcommittee members that the utilities
were "fully committed to supporting additional research"
into the biological effects of electromagnetic fields, but that
there was presently "insufficient information to justify any
change in utility operations." In this connection, he said
that even burying power lines might not reduce exposure

to electromagnetic fields. (Apparently, he did not know of the 1989 study that had been prepared by eight major utilities in New York State, and coordinated through the Empire State Electric Energy Research Corporation; it showed that when a three-hundred-and-forty-five-thousand-volt transmission line is placed in steel pipe filled with oil and buried at a depth of five feet, the strength of the magnetic field measured one meter above the ground over the pipe is only about one milligauss during normal current flow, as compared with magnetic-field strengths as high as sixty milligauss that can be measured at the edge of a one-hundred-foot right-of-way for an overhead three-hundred-and-forty-five-thousand-volt transmission line.) Zweiacker went on to say that "a definitive conclusion that EMF is harmful cannot be drawn solely from epidemiological studies," and that since "no study has identified what kind of exposure may have adverse health effects, it is virtually impossible to mitigate EMF appropriately, even if there is a problem."

During the summer, E.P.A. officials revised the draft report a second time, before submitting it for review to the Agency's Scientific Advisory Board. In September, the Advisory Board began selecting a seventeen-member subcommittee on electric and magnetic fields, and named Dr. Genevieve Matanoski, of Johns Hopkins University, to head it.

Meanwhile, in July, William Reilly, the administrator of the E.P.A., had asked D. Allan Bromley, a former professor of physics at Yale University, who was the director of the White House Office of Science and Technology Policy and the chief science adviser to President Bush, to arrange for an additional review of the report, to be carried out by the Committee on Interagency Radiation Research and Policy Coordination (CIRRPC), which is an offshoot of the Office of Science and Technology Policy. In August, Bromley not

only arranged for the review, but he also sent Reilly a list
of fourteen people whom he recommended for inclusion
on the E.P.A. Scientific Advisory Board's subcommittee, and
whom he described as being "very knowledgeable" about
the biological effects of power-line electromagnetic fields,
and thus capable of assisting Reilly in assessing "a very difficult area that I am sure you agree requires careful treatment if we are to serve the public well."

Coming from the nation's highest-ranking science-policy
official, Bromley's list was disingenuous at best and insensitive to conflict of interest at worst. Of the fourteen experts
he recommended for Reilly's consideration, four were paid
consultants of the electric utility industry, who had testified
in behalf of the industry in court cases involving the health
hazard posed by power-line electromagnetic fields; three
others had published articles suggesting that further research on the biological effects of low-level electromagnetic radiation be suspended; two worked for the Naval
Aerospace Medical Research Laboratory, an outfit that during the nineteen-seventies had conducted human experiments showing that extra-low-frequency fields could have
significant effects upon behavior; two had been financed by
the Air Force, which had been trying to suppress information about the adverse health effects of low-level electromagnetic radiation since the early nineteen-sixties; and one
worked for the General Electric Company, a firm that had
denied the possibility of adverse health effects from such
radiation for an equal period of time. The remaining two —
Robert K. Adair, a professor of physics at Yale University and
a former colleague of Bromley, and his wife, Eleanor R.
Adair, who is a fellow of the John B. Pierce Foundation Laboratory at Yale's Center for Research in Health and the Environment — had stated publicly on a number of occasions

that the nationwide concern about power-line magnetic fields was nothing more than mass hysteria.

To their credit, Reilly and Donald Barnes, the director of the E.P.A.'s Scientific Advisory Board, did not select any of the people whom Bromley had suggested as members of the subcommittee on electric and magnetic fields. However, of the seventeen members of the panel that was selected six members either had been paid consultants of the utility industry or had conducted studies financed by its research arm — the Electric Power Research Institute, or EPRI, of Palo Alto, California. By this time, Reilly and other high Agency officials were aware that powerful forces within the government were prepared to oppose any attempt on the part of the E.P.A. to characterize low-level electromagnetic fields as a possible cancer hazard. On October 2nd, Brigadier General Paul D. Gleason, director of professional affairs and quality assurance for the Air Force's Office of the Surgeon General, sent a letter to Erich Bretthauer, the E.P.A. assistant administrator for research and development, warning Bretthauer that if the E.P.A. draft report were published it would "contribute to needless public anxiety, and have serious impacts on capabilities and costs of Air Force programs." (Mounting public opposition to the Ground Wave Emergency Network [GWEN] — a communications system consisting of ninety-six very-low-frequency radio transmitters that the Air Force had proposed to install across the continental United States, in order to withstand the disruptive electromagnetic pulse that would accompany the detonations of a nuclear attack — had recently forced Air Force officials to ask the National Academy of Science's National Research Council to review the possible health hazard posed by exposure of people in the general population to radiation emitted by the GWEN transmitters.) Gleason told Bretthauer that the consequences of such anxiety, in view of the "minor nature" of the biological effects cited in the report, "are especially unwarranted." He went on to say that the Air Force opposed publication of the E.P.A. report until assessments of the current state of science regarding the biological effects of electromagnetic radiation had been completed by the National Academy of Sciences and the Committee on Interagency Radiation Research and Policy Coordination.

Gleason's letter was accompanied by a harshly worded thirty-three-page critique of the E.P.A. report, which had been compiled by officials of the Air Force's school of Aerospace Medicine, at Brooks Air Force Base, in Texas — an organization that for more than two decades had vigorously opposed any suggestion that the electromagnetic radiation emitted by Air Force radar and communications equipment could be hazardous. The authors of the Air Force critique declared that the scientists who had written the E.P.A. report "have biased the entire document" by making "political statements and not scientifically derived conclusions." They went on to say that "the overall weight of epidemiologic evidence is so slight as to be almost non-existent," and to claim that there was "no suggestion that EMFs present in the environment today induce or promote cancer." They concluded their critique by voicing concern that release of the Agency's report would cause "an epidemic of apprehension."

Apparently, neither Bretthauer nor Reilly saw fit to question the propriety of an Air Force general characterizing excess cancer in children as a biological effect that was minor in nature, or, for that matter, the propriety of the Air Force involving itself in a public health matter affecting the civilian population of the United States. In any event, they soon encountered more formidable opposition to the release of the draft report in the person of Bromley. On November 26th, Bromley was briefed at the White House by Bretthauer, Farland, and Robert E. McGaughy, the project manager and chief author of the report. Bromley then told the three men that the report would alarm the public and asked them to delay its release until it could be further evaluated.

The report had been scheduled for release on November 27th, and when it was held up because of Bromley's request, several junior officials at the Agency notified the Associated Press. A story by Paul Raeburn that went out on the AP wire on December 13th quoted McGaughy as saying that Bromley had voiced concern that the report would alarm the public, and that similar reservations had been expressed by Assistant Secretary of Health James Mason a week earlier. "They were concerned not about the accuracy of the report," McGaughy told Raeburn. "They were concerned about how people would react to the news."

That evening, two major networks broadcast stories based on the AP account, and the next day, December 14th, various versions of the AP account, or stories based on it, were carried by newspapers across the nation. Among them were *USA Today, New York Newsday,* the *Houston Post,* the *Wisconsin State Journal,* and the *Philadelphia Inquirer.* The *Fresno Bee* carried the AP story in its entirety. That same day, the E.P.A. released the report, but with a disclaimer that mentioned the "controversial and uncertain nature of the scientific findings of this report," and declared that it "should not be considered as representing Agency policy or position."

The AP account and most of the newspaper stories that followed emphasized the Administration's attempt to delay the release of the E.P.A. report, but told their readers very little about its findings. For example, few of the stories mentioned the fact that two studies cited in the E.P.A. report had shown that children exposed to power-line magnetic fields at or above two to three milligauss were experiencing significantly elevated risks of developing leukemia, brain cancer, lymphoma, or other malignancies, let alone the fact that magnetic fields of such strength could be measured routinely in dwellings and schools across the nation that are situated close to high-voltage transmission or high-current distribution lines. Few of the accounts took note of the E.P.A.'s conclusion that the findings of the childhood studies were supported by the findings of other studies showing that men engaged in electrical and electronic occupations were experiencing a significantly increased risk of developing leukemia, brain cancer, and malignant melanoma than other workers. Moreover, none of the stories mentioned the Agency's judgment that the results of laboratory studies showing changes in brain chemistry, impairment of the immune system, and inhibition of melatonin synthesis by the pineal gland lent plausibility to the epidemiological evidence that power-line magnetic fields could be carcinogenic in humans.

On December 15th, the *New York Times* ran a story by Philip J. Hilts, in which Bretthauer was quoted as saying that "no responsible scientist can make a conclusion that there is a cause and effect" between electromagnetic fields and cancer or other disorders. Bretthauer went on to say that even if further research did bear out the association between electromagnetic fields and cancer, it would probably be years before the findings were certain enough to result in federal regulatory action. Farland also tried to cast doubt on the importance of the report by telling Hilts that the biological effects of electromagnetic fields could not amount to much, because the records of disease in this century did not show notable increases as the electrification of the nation progressed. Only four days earlier, the *Times* had carried a story about steep increases that had been observed in the United States and elsewhere in the world in the incidence of brain cancer, lymphoma, and melanoma — all of them diseases that had been cited in the report compiled by scientists in Farland's own office as being associated with exposure to power-line magnetic fields. The earlier *Times* story noted that scientists had implicated electromagnetic radiation as a possible cause of brain cancer.

Meanwhile, Representative George E. Brown, of California, the chairman of the House Committee on Science, Space, and Technology, together with Representative James H. Scheuer, of New York, the chairman of that committee's subcommittee on Natural Resources, Agriculture Research, and Environment, and Representative Frank Pallone, Jr., of New Jersey, the author of a bill seeking more federal funding for a study of the biological effects of electromagnetic fields, had written Bromley expressing dismay over his role in delaying the release of the E.P.A. report, and to tell him that his action was "more likely to fan public concern than to allay it." Bromley, in a letter dated December 17th, told the three congressmen that a number of his "physicist friends," including Robert L. Park, of the American Physical Society, had "serious questions about the quality of the studies" that had been conducted thus far on the suspected carcinogenicity of electromagnetic fields.

Bromley's regard for Park's views was revealing in that Park, who is executive director of the American Physical Society's Office of Public Affairs and the chairman of the physics department at the University of Maryland, not only entertained serious questions about the carcinogenicity of electromagnetic fields, but had also seen fit to make light of the problem. "The market for fear has never been better," he wrote in an opinion piece that appeared in the Sunday edition of *Newsday* on October 29, 1989. "The sky rains acid on dying lakes and forests; the ocean regurgitates the filth from our cities; toxic fumes percolate up from forgotten waste dumps beneath our homes; lead is in our water and Alar on our apples; we are being smothered by a thickening blanket of carbon dioxide, while chlorofluorocarbons are eating a hole in the ozone. And now we are told that the fields produced by ordinary household electricity may be addling our brains and inducing leukemia in our children." Park tried to play down the power-line hazard by pointing out that life expectancy in the United States had nearly doubled since the Industrial Revolution began, and that the increase had been most rapid since the introduction of electricity. (In the summer of 1990, Eleanor Adair would echo him by telling the *New York Times* that "electricity has been around for a long, long time, and people's life expectancy has nearly doubled since the invention of the light bulb.") After comparing the repeated associations that had been found between exposure to electromagnetic fields and the development of cancer with the false claims surrounding the alleged discovery of cold fusion, Park declared that there was no known biological response to electromagnetic fields that would "lead one to expect harmful effects," and that there were "at most a few contradictory reports of weak biological responses." He went on to say that living organisms respond to "all sorts of stimuli," and that "it is possible to measure a rather strong physiological response in humans to the smell of freshly baked bread, but no one suggests it is harmful."

As for what had occurred at the November 26th meeting,

Bromley told the three congressmen in his letter that he had not tried to censor the E.P.A. report but that he considered it his responsibility to urge caution "when a very large population stands to be affected negatively by the report." On December 24th, Bromley was quoted in *Time* as saying that the EPA's finding of a positive association between exposure to electromagnetic fields and childhood cancer was "unnecessarily frightening millions of parents."

For his part, Reilly, the E.P.A. administrator, appeared to be condoning the actions of subordinates, who were paving the way for the utility industry to mount an all-out attack on the E.P.A. report and its conclusions at the initial Science Advisory Board hearings, which had been scheduled for mid-January of 1991. In a letter sent to Reilly on December 21st, Representatives Brown, Scheuer, and Pallone sharply criticized the E.P.A. for allowing Crowell & Moring — a Washington law firm that represented many utility companies, including several that had been named as defendants in lawsuits involving the health hazards of power-line magnetic fields — to schedule public comment at the upcoming hearings on key issues, such as epidemiology, cell biology, and biophysics. The three congressmen told Reilly that this "suggests an agenda which will receive disproportionate input from the client interests of Crowell & Moring." (By then, many of the slots for public comment had already been assigned to speakers with ties to the utility industry.) The congressmen also pointed out that public notice of the hearings had not appeared in the *Federal Register* until December 18th, and that because Crowell & Moring had received advance notice of them it enjoyed a "significant advantage" over other interested parties. "The very tradition of scientific peer review is founded on the notion that truth will emerge from rigorous debate and a close examination of all viewpoints," they wrote. "To give a party with such clear vested interests as Crowell & Moring the power to 'arrange' the testimony before the Science Advisory Board is totally contrary to this tradition. If nothing else, the 'stacked deck' appearance of the presentations will destroy the very credibility of the SAB review process."

As it happened, this was not the first time that Crowell & Moring had tried to stack the deck in behalf of a utility industry client. In 1987 and 1988, the firm had succeeded in hiring three high-ranking scientists from the National Cancer Institute to testify for the New York Power Authority in a sixty-six-million-dollar lawsuit that had been brought by landowners, who alleged that a two-hundred-and-seven-mile-long, three-hundred-and-forty-five-thousand-volt transmission line that the Authority had built through several counties in upstate New York posed a cancer hazard and had destroyed the market value of property adjacent to its right-of-way. The three NCI scientists, who had never conducted any research of their own into the biological effects of power-line electromagnetic fields, were paid a total of more than $124,000 for testifying that there was no scientific basis upon which to conclude that the fields could cause cancer. The plaintiffs lost the case when the judge who was hearing it sided with the witnesses who had been assembled by Crowell & Moring. As for the three NCI scientists, they were subsequently found to have violated rules governing outside income that could be earned by employees of the National Institutes of Health by accepting unauthorized fees.

In preparation for the hearings, Crowell & Moring had set up an organization called the Utility Health Sciences Group — a coalition of major utility companies that claimed to be interested in promoting research on electromagnetic fields — and the group had arranged for four prominent scientists to come to Washington to give testimony before the S.A.B. subcommittee which would discount the association between electromagnetic fields and cancer. The four were Dr. David Korn, the dean of Stanford University's School of Medicine, who was also chairman of the National Cancer Advisory Board; Dr. Mark Mandelkern, a professor of physics and pathology at the University of California, Irvine; Dr. Edward Gelmann, the chief of medical oncology at the Georgetown University School of Medicine's Vincent Lombardi Cancer Research Center; and Dr. Dimitrios Trichopoulos, the chairman of the Department of Epidemiology of Harvard University's School of Public Health.

Dr. Korn testified that on the basis of his readings and inquiries "the case for the potential carcinogenicity of power frequency electromagnetic fields is not convincing," and that the evidence to date was "vastly insufficient to support any kind of sound decision-making with respect to new cancer regulatory policy." However, when he was questioned by members of the subcommittee, he admitted that he had not read any of the key papers on the cellular and animal effects of electromagnetic fields. Dr. Mandelkern declared that the mechanisms of interaction between electromagnetic fields and biological processes were "implausible," but, like Korn, he was also shown by questioners to be unfamiliar with much of the literature on the effects of low-level electromagnetic fields. Dr. Gelmann stated that "there appears to be no substantive evidence that power-frequency electric and/or magnetic fields induce cancer." Dr. Trichopoulos, echoing Farland, suggested that if the proposed association between exposure to electromagnetic fields and cancer were true, the increasing electrification of the nation over the years should have resulted in an "epidemic" of childhood leukemia. This prompted Dr. Matanoski, who was chairing the S.A.B. subcommittee, to point out that there had been a consistent increase in childhood leukemia since the 1930s. She was correcting a canard that had been floating about in utility-industry public relations circles since the summer of 1989, when a member of Pacific Gas & Electric's task force on electromagnetic fields had tried to allay concern about the studies linking power-line emissions to childhood leukemia by writing in the company's weekly newsletter that "while use of electricity in this country has increased fivefold in the last 30 years, the reported number of new leukemia cases has fallen by about half."

During the hearings, eighteen oral comments were given, and more than half of them were delivered by people who represented the utility industry or the Air Force or who had publicly stated that there was no evidence of an association between exposure to electromagnetic fields and cancer. Among the latter was Robert Adair, who had recently told a reporter for the *Los Angeles Times* that investigating the biological effects of electromagnetic fields was like looking for werewolves. Adair told the subcommittee that the chapter in the EPA report on mechanisms of interaction was "crackpot science." He went on to claim that universally accepted laws of physics made it virtually impossible that sixty-hertz magnetic fields at levels below a hundred milligauss could cause any observable biological effects. Moreover, there were only half a dozen comments by people who were concerned that power-line fields might pose a health risk, and not one was delivered by anyone with a scientific research background, let alone by any of the scientists who had conducted any of the dozens of studies that had been cited in the E.P.A. report as showing that electromagnetic fields were a possible cause of cancer. The resulting imbalance was noted by Mayor James P. Connors, of Scranton, Pennsylvania, who told the sub-committee that many of his constituents were worried about their exposure to power-line emissions. (Some Scranton residents suspected that several cases of cancer and Hodgkin's disease within a single neighborhood had been caused by a nearby sixty-nine-thousand-volt power line that was carrying high current and giving off strong magnetic fields.) After pointing out that a large number of speakers at the hearings represented the utility industry, Connors asked, "Who is representing the people?"

CHAPTER NINE

Consider That
the Roosters Crow

MEANWHILE, in Connecticut, there had been renewed calls for statewide studies to determine the extent of the power-line hazard, and whether brain tumors were developing more readily in people exposed to power-line magnetic fields. Additional impetus for such studies came when Don Michak reported that he and some colleagues at the Manchester *Journal Inquirer* had measured power-line magnetic fields near a number of schools and found them to be in excess of the two-to-three-milligauss level that had already been associated with an increased incidence of cancer in children. Among their readings were fields of more than four milligauss in front of a day-care facility in Windsor Locks; fields of up to seven milligauss on the sliding board of a playground at the Maple Street School, in Vernon; and a field of nearly eight milligauss at the entrance to St. Joseph's Elementary School, which is also located in Vernon. A day after the *Journal Inquirer* article appeared, a spokesman for Northeast Utilities declared that such magnetic-field levels fell within a "normal range." He went on to say that a brain tumor study in Connecticut would be too costly for the deficit-ridden state government to undertake.

On November 23rd, 1990, Senator Cornelius P. O'Leary, a Democrat from Windsor Locks, the majority leader of the Connecticut State Senate, announced that he and other leading senators — among them the chairmen of the Senate Energy and Public Utilities Committee, the Public Safety Committee, and the Environment Committee — would ask the Connecticut Academy of Science and Engineering to conduct a study of the health effects of power-line magnetic fields and report its findings to the General Assembly. (The Academy is a nonprofit corporation that had been formed by the legislature in 1976 to "promote the application of science and engineering to human health and welfare.") State Health Commissioner Frederick G. Adams joined O'Leary and his Senate colleagues in calling for the study, and asked the Academy to conduct an independent review of the Department of Health Services' conclusion that there was no evidence of a cancer cluster on Meadow Street. In the meantime, Adams advised Connecticut residents to reduce their exposure to electromagnetic fields until more conclusive evidence about the potential health hazard became available. Senator Richard Blumenthal, a Democrat from Stamford, who had recently won election as Connecticut's next attorney general, declared that a statewide study of the health effects of power-line electromagnetic fields should be among the top items on the agenda of incoming Governor Lowell P. Weicker, Jr.'s new administration. In the middle of December, O'Leary wrote to the members of a thirteen-member interagency EMF task force that had been set up to study the power-line problem, asking them if it was appropriate to propose "prudent avoidance" measures that the General Assembly might enact into legislation during its next session. The task force had been formed earlier in response to criticism that state health officials had failed to properly investigate the power-line health hazard. Its members included the Department of Health Services, the Department of Environmental Protection, the Department of Public Utility Control, the Connecticut Siting Council, the Department of Consumer Protection, the Department of Economic Development, and the Office of Policy and Management.

The calls for action by O'Leary, Adams, and Blumenthal coincided with nationwide publicity that had accompanied the attempt by officials of the Bush Administration to delay publication of the Environmental Protection Agency's draft report on the carcinogenicity of power-frequency magnetic fields. Meanwhile, doubts about the fitness of the Connecticut Academy of Science and Engineering to conduct an unbiased study of the power-line hazard were raised by Michak, who reported in the *Journal Inquirer* that the Academy had received financial contributions from Northeast Utilities and United Illuminating, of New Haven — the two largest utilities in the state — and that among the Academy's other corporate sponsors were the General Electric Company and Combustion Engineering, Inc., both of which are large manufacturers of electrical generating equipment, as well as the Southern New England Telephone Company, whose financial interests could also be affected by the Academy's proposed study of power-line magnetic fields. Michak also revealed that the director of environmental programs for Northeast Utilities was a member of the Academy's board of directors, and that the company's vice-president for corporate planning and regulatory relations was chairman of the Academy's nominating committee. At the time the story appeared, officials of the Academy told

Michak that any Academy members whose corporate associations might prejudice the outcome of the study would
not be allowed to participate in it.

On January 23rd, 1991, Dr. Peter Galbraith, of the Department of Health Services, and Carl S. Pavetto, chief of the
Department of Environmental Protection's air management
bureau, who had been named co-chairmen of the interagency task force, wrote to Senator O'Leary regarding his
earlier request that the task force recommend prudent-
avoidance measures for the General Assembly to consider.
The two officials told O'Leary that "we don't have enough
information to be able to say with certainty that we should
recommend prudent avoidance," and that they preferred to
wait for the results of ongoing studies of the power-line
hazard before recommending any legislative action. A few
days later, David R. Brown and two of his colleagues in the
Division of Environmental Epidemiology and Occupational
Health wrote to Suzanne Bullock, Melissa Bullock's mother,
and other residents of Meadow Street, assuring them that
"we are continuing to actively follow your concerns about
health and electromagnetic fields." The three health officials went on to say that "it is important to note that an association between residential proximity to power lines and
cancer does not mean that the fields from the power line
cause cancer." They then advised Mrs. Bullock and the
other residents of Meadow Street to reflect upon how mis-
conceptions about the carcinogenicity of power-line magnetic fields can arise, using an analogy borrowed from a
pamphlet that had been written by researchers at Carnegie
Mellon University, in Pittsburgh. "Consider as an example
that the temperature rises each morning after the roosters
crow," they wrote. "While these two factors are associated
with each other, we know that roosters do not cause the

temperature to rise. Rather, sunrise causes both of these phenomena."

Similar efforts to play down the power-line hazard would soon be made by utility officials and state health authorities in California. On February 25, 1991, Thom DeYoung, a service planning analyst for the Pacific Gas and Electric Company, sent a thirty-two-page report of the magnetic-field measurements the utility had taken at the Lawless and Slater schools in December, together with some background material on the electromagnetic field hazard, to Wayne McMillen, the director of the Fresno Unified School District's Benefit and Risk Management Department. In a cover letter, DeYoung told McMillen that the utility was "unable to interpret" the measurements, and that the lack of scientifically based safety standards for human exposure to power-frequency magnetic fields meant that individuals had to formulate their own risk assessment. The report noted that P. G. & E.'s facilities near the Slater Elementary School included a two-hundred-and-thirty-thousand-volt double-circuit transmission line and a hundred-and-fifteen-thousand-volt double-circuit transmission line, and that the latter was located "approximately 100 feet from the nearest classrooms." It went on to say that between three and four o'clock on the afternoon of December 17, 1990, a magnetic-field level of 7.6 milligauss had been measured directly beneath the hundred-and-fifteen-thousand-volt line on Emerson Avenue; a level of 5.2 milligauss had been found at the curb on the school side of the avenue; a level of 4.3 milligauss had been found by a fence separating the school grounds from the sidewalk by the avenue; and levels of between one and two milligauss had been measured at nine locations immediately outside Pods A and B.

On March 1, McMillen forwarded a copy of the P. G. & E. report to Chuck McAlexander, an assistant superintendent of the school district, with a cover letter assuring McAlexander that not enough scientific information existed to make a judgment regarding exposure levels. To understand how McMillen may have come to this conclusion one need only read the background material that DeYoung had sent him. It included a public-policy statement that had been published by P. G. & E. in October of 1990 and a pamphlet that had been written earlier that year by Dr. Raymond Richard Neutra, the chief of the California Department of Health Services' Special Epidemiological Studies Program, and three of his colleagues, who supported the utility's assessment at almost every turn.

Neutra and his co-authors started out by declaring that "the popular press has recently focused public attention on the scientific controversy as to whether there are any health effects from magnetic and electric fields of power lines and appliances," and that "as a result, utility companies and public health agencies have been receiving calls from concerned homeowners requesting that magnetic field measurements be made and that the health implications of these measurements be given." Of course, the chief reason that the press had focused public attention on the power-line hazard was that the E.P.A. report had described power-line magnetic fields as first a "probable" and then a "possible" cause of cancer in humans. (The fact that Neutra was familiar with the E.P.A. report is indisputable; he had been one of nine scientists on an E.P.A. review panel that had assessed the report in June of 1990, and had praised it as an "excellent" document in a letter he wrote to Robert McGaughy, its chief author, on July 9th.) Neutra and his colleagues went on to say that, given the scientific information now available, it was not possible to set a standard for exposure to electromagnetic fields or to say that any given level was either safe or dangerous. They advised that "individuals who are concerned may choose to take steps such as moving an electric clock a few feet away from a bedside table or working on their computer key board further away from a screen, or perhaps not using some electrical appliances at all." They had no advice for individuals such as the teachers and parents of children at the Slater Elementary School, who might be concerned because the school was situated very close to a pair of high-voltage transmission lines, which give off stronger fields at greater distances than any appliance, and, by subjecting people within their range to long-term exposure, pose a far more serious health risk.

In a later section of the pamphlet, Neutra and his colleagues implied that epidemiological studies with findings that supported a link between exposure to power-line magnetic fields and the development of cancer were balanced by a similar number of studies with findings that did not support such a link. The fact is that, as the E.P.A. draft report clearly stated, five out of six studies of childhood residential exposure and most of thirty or so occupational studies that had been conducted since 1979 showed a positive link between exposure to power-frequency magnetic fields and cancer. Neutra and his co-authors went on to say that the twofold increased risk of cancer that had been observed in children who lived in homes close to high-current distribution lines was "similar to the increased lung cancer risk conveyed to an adult non-smoker from living in the household of someone who smokes." This analogy was inappropriate for two reasons: first, because children exposed to power-line magnetic fields have been shown to develop leukemia, brain cancer, and lymphoma, not lung cancer; and, secondly, because an adult non-smoker living in the household of someone who smokes has the opportunity to consent or not to consent to his or her exposure to benzopyrene, the carcinogen in tobacco, whereas a child living in a home or attending a school close to power lines giving off biologically significant magnetic fields would be undergoing involuntary exposure to a potential carcinogen.

Neutra and his colleagues also said that to explain the findings of the epidemiological studies "we must accept an entirely new way for very weak magnetic fields to affect molecules in the body and a new pathway for such changes to cause cancer or other health effects." They said almost nothing about the research demonstrating that electromagnetic fields could alter the chemistry of the brain, impair the immune system, and inhibit the synthesis of melatonin, which had provided the chief basis for the E.P.A.'s conclusion that the findings of carcinogenicity were "biologically plausible." However, they declared that "scientists usually require very strong evidence before they will adopt a radical new theory," and that "the fact that any recommendations for 'safe levels' might have to be revised after widespread and costly efforts had started, requires scientists to really understand the problem before giving advice." This implied that public health researchers should refrain from recommending interim preventive measures to reduce people's exposure to power-line magnetic fields until they understood exactly how magnetic fields might operate to cause cancer — an absurdly paralyzing requirement that would have prevented health scientists, who still do not understand how inhaled asbestos fibers react to cause cancer in lung tissue, from having recommended that asbestos be removed from the nation's schools.

Neutra and his colleagues emphasized the fact that the childhood epidemiological studies had relied on rough estimates of magnetic-field exposure based on the proximity of homes to power lines, rather than on spot measurements of actual magnetic-field strength, but they failed to point out that most leading researchers consider such estimates to be historically more accurate and to give a better idea of actual exposure at the time of and before diagnosis than spot measurements. They went on to say that "we do not know (as of this writing in 1990) which aspects of 60 Hz fields in a home should be measured or how exposure levels should be managed, if at all." Having thrown up their hands about the power-line problem, they said that "there are some who feel that carrying out magnetic field measurements in the face of such uncertainties serves no scientific purpose, and will just cause people to worry unduly" — a position that had been taken by many of the nation's utilities — but they conceded that "the inability to obtain even the simplest measurements about one's home and school environment may be frustrating and anxiety-provoking in and of itself." They concluded their pamphlet by stating that public concern about the power-line health hazard was based on "incomplete and inconclusive" data, and that until better data were available concerned individuals might wish to limit their exposures to electromagnetic fields, "when this can be done at reasonable cost and with reasonable effort."

Like Neutra and his colleagues, the authors of the P. G. & E. statement emphasized the fact that estimates of magnetic-field exposure in the childhood-cancer studies had been based on proximity to power lines and not on actual magnetic-field measurements, and thus did not prove a link between exposure to electromagnetic fields and the development of cancer. They also pointed out that "current scientific knowledge cannot explain how the extremely weak electric and magnetic fields associated with power lines might produce human health effects, including leukemia." In addition, like Farland and Trichopoulos, they suggested that if power-line magnetic fields really caused or promoted cancer, "the fivefold increase in electrical usage during the past 30 years would have been expected to have produced an epidemic of childhood leukemia." The utility industry stopped making this argument in July of 1991, after the National Cancer Institute released a report saying that there had been unexplained increases in recent years of nearly twenty-five per cent in childhood leukemia, and of more than thirty percent in childhood brain cancer. The authors of the P. G. & E. statement went on to say that the utility was "working with State and local agencies to provide up-to-date information about the issue to the agencies and the public in general," and that it was participating in a California Department of Health Services' effort to develop a measuring procedure and communications program.

Meanwhile, other utilities in California were taking positions on the electromagnetic-field hazard that were similar to those taken by P. G. & E. A two-page brochure being mailed out by Southern California Edison to citizens who inquired about the power-line hazard claimed that Edison was "working with the California Department of Health Services and other government agencies to resolve your concern and share all new information as it is developed." The brochure quoted a 1989 report that had been issued jointly by officials of the Department of Health Services and the California Public Utilities Commission, who recommended that the state "take no action at present to regulate electric and magnetic fields around electric power facilities," and declared that "such actions are premature given current scientific understanding of this public health issue."

A ten-page pamphlet that had been issued by San Diego Gas & Electric in November of 1990 described how Neutra regarded the hazard posed by power lines. The S.D.G. & E. pamphlet quoted him as saying that although no one knew for sure if there were any harmful effects from electromagnetic fields, people might wish to reduce their exposure "without causing a lot of trouble to society" by moving their electric clock radios, or by maintaining a greater distance from their video display terminals. "When you talk about schools, moving kids, selling homes, moving power lines, we are not advocating anything like that," Neutra said. "We don't think the evidence now supports that kind of drastic action. It may turn out some of those things are needed, but you have to build a case."

CHAPTER TEN

Some Red Dots
on a Map

W HILE UTILITY officials and state health authorities in California were extending the benefit of the doubt to the power lines, evidence about the cancer hazard associated with power-line emissions was growing by leaps and bounds. Although some of this new evidence had profoundly troubling implications for the future health and well-being of the teachers and children at Slater, neither the teachers nor the parents of the children nor, for that matter, anyone else in Fresno was told anything about it by P. G. & E. or the Department of Health Services, in spite of the utility's assertion that it was working closely with the Department to provide the public with "up-to-date information" about the power-line health issue. On January 30th and 31st of 1991, a panel of epidemiologists who were attending a workshop on the health effects of electromagnetic radiation that was being held in Cincinnati by the National Institute for Occupational Safety and Health strongly urged that the association of female breast cancer and exposure to electromagnetic fields be given high priority in future research programs. Just a month earlier, a

Norwegian study had been published which showed that male breast cancer — a disease so rare that only about one case is diagnosed in every one hundred thousand men each year — was occurring at twice the expected rate in men whose occupations involved exposure to electromagnetic fields, and at four times the expected rate in electric transport workers, such as railway engine drivers, tram operators, and track walkers.

The Norwegian study was the third investigation in little more than a year to observe excess breast cancer among men exposed to electromagnetic fields. Seven months earlier, epidemiologists at the Fred Hutchinson Cancer Research Center, in Seattle, had reported that telephone linemen, electricians, and electric power workmen were developing breast cancer at six times the expected rate, and in November of 1989 Matanoski and her colleagues at Johns Hopkins had announced that there was an increased incidence of breast cancer among central-office telephone technicians, who were exposed to magnetic fields of between two and three milligauss being given off by switching equipment. (The increase also proved to be six times as great as expected.) In spite of the fact that Louis Slesin had reported all these findings in *Microwave News,* almost no word of them had found its way into the nation's newspapers, with the result that American women, who were experiencing a steep rise in the incidence of breast cancer — a malignancy that is histologically identical to breast cancer in men — had almost no idea of the potential link that had been discovered between the development of this disease and exposure to magnetic fields.

Among the epidemiologists at the Cincinnati workshop who stressed the importance of conducting studies to determine the possible association between female breast

cancer and electromagnetic fields was Dr. John M. Peters, of
the Department of Preventive Medicine of the University of
Southern California School of Medicine, in Los Angeles. Peters and Dr. Stephanie J. London, together with some
colleagues at the university, had recently completed a five-
year investigation of the association between childhood
leukemia and residential proximity to high-current power
lines — an investigation financed by the Electric Power Research Institute. On February 7th, EPRI officials announced
the preliminary results of the study at a meeting held in
Carmel, California, which was closed to the press. According to EPRI, London, Peters, and the other U.S.C. researchers had examined the residences of two hundred and
thirty-two children, ten years old or younger, who lived in
Los Angeles County and had developed leukemia between
1980 and 1987. When they compared these residences with
those of two hundred and thirty-two children who had not
been afflicted with leukemia, they found that children living near high-current power lines experienced a risk of developing the disease two-and-a-half-times as great as
children who did not live near such lines, and thus confirmed the findings of earlier childhood cancer studies that
had been cited in the E.P.A. report as providing the
strongest evidence that power-line magnetic fields were a
possible cause of cancer in human beings.

Unlike previous investigators, the U.S.C. researchers
took magnetic-field readings of at least twenty-four-hours
duration in the bedrooms of three hundred and five out of
six hundred and ninety-two of the residences that had been
lived in by the four hundred and sixty-four children in their
study and for which they had been able to obtain wire-code
configurations. The fifty per cent increased risk of leukemia
they observed in the highest exposure category — children

in whose bedrooms magnetic fields of two and two-thirds milligauss or above were recorded — was not considered to be statistically significant. Since they had decided not to engage in any public discussion of their study until it was published in a peer-reviewed scientific journal — and that did not occur for nine months — EPRI was able to put its spin on the findings by emphasizing, in a seven-page commentary it distributed to the press, that measured magnetic-field exposures were only "weakly associated with leukemia risk," and that the "lack of a clear pattern of association with different measures of exposure requires further research." In spite of this, the *Wall Street Journal, Los Angeles Times,* and most of the other major newspapers that covered the story placed the London-Peters study in proper perspective by reporting that children living near high-current power lines had been found to be developing leukemia at a significantly higher rate than other children, and that this finding constituted important new evidence of the link between exposure to electromagnetic fields and cancer. An exception was the *New York Times,* whose Sandra Blakeslee wrote a lead sentence declaring, "Preliminary results of a new scientific study show that childhood leukemia is not associated with household exposure to electromagnetic fields." No mention of the London-Peters study appeared in the *Fresno Bee,* and therefore neither Patricia Berryman and her fellow teachers at the Slater School nor the parents of the children who attended the school were aware that the findings of a major new investigation suggested that it was hazardous for children to be chronically exposed to power-line magnetic fields.

A few days after Wayne McMillen forwarded P. G. & E.'s report to McAlexander, Berryman picked up a copy from a table in the faculty room at the Slater School, where George

Marsh, Slater's principal, had left it. Reading through it, she found that it consisted largely of a series of graphs showing magnetic-field profiles that had been taken at specific locations to substantiate the measurements in the graphs. She also found that all of the measurements were expressed in terms of milligauss, which made little sense to her, because at that time she had no idea what a milligauss signified. For this reason, she was unable to attach any significance to the ambient magnetic-field readings that P. G. & E. had taken near the classrooms of Pods A and B. On page 4, however, she came across a description of the measurements which caught her attention. It stated, "Generally, the higher ambient magnetic fields occur on the south side of the school property (in the vicinity of Pods A and B) due to the proximity of the power lines."

At that time, faculty meetings were held at Slater on the first two Tuesdays of each month, and when the P. G. & E. report came up for discussion on the afternoon of March 5th, Berryman told her colleagues that she had not been able to understand much of it. When a similar reaction was expressed by Curtis Hurd, one of the vice-principals, George Marsh announced that P. G. & E. would be presenting the results of its survey at a meeting of the Fresno Board of Education, on March 14th, and suggested that anyone wishing further clarification should attend. Discussion among the teachers turned to the article about the power-line hazard that had run in December in the *Bee;* to the Associated Press story about the EPA report and its mention of a link between exposure to power-line magnetic fields and brain cancer; to the fact that Katie Alexander, who had worked in Pod A, had developed brain cancer, and that another teacher, who had worked in Pod B, had been disabled by a brain tumor; and, finally, to the general awareness among them that all the cancers and tumors that had been diagnosed in women working at Slater over the previous eight years had occurred among women who had worked on the south side of the school — either in Pods A and B, or in the adjacent administrative areas — near the high-voltage transmission lines on Emerson Avenue.

When the faculty meeting adjourned, Berryman and several of her colleagues stayed behind to discuss the matter further, and decided that they should send a representative to the March 14th Board of Education meeting. Berryman then suggested that they draw up a list of the names of everyone who had developed cancer while working in or near Pods A and B, and make a diagram showing the locations in which they had worked, so that the members of the Board would be able to see that, as Berryman put it, "we have had a lot of sick teachers at this school." Doris Buffo, a fifth- and sixth-grade teacher, who had worked in Pod B for ten years, remembers that someone drew a crude sketch of the school and marked it with "Xs" to show where the cancer victims had worked. "When I looked at all those 'Xs,' prickles went up and down my spine," she said recently. "Back in December, when I had read about the power lines in the *Bee,* I had thought it was just another one of those you-can't-breathe-the-air-or-drink-the-water-and-now-you've-got-something-new-to-worry-about   kinds of things. But all of a sudden I realized that I had spent ten years in Pod B, and I thought, 'My God, this is scary!' "

That afternoon, Buffo went to Marsh's office and obtained a small map of the Slater School. She enlarged it with an overhead projector in her classroom and traced the outline of the enlargement on a two-by-three-foot piece of poster board. At the end of the day, she took the poster board home and affixed red circular stickers to it to mark the workplaces of cancer victims. When she finished, there were seven red circles at various locations in Pods A and B, and three circles in the administrative offices just behind Pod A. The next morning, she took the diagram to school and showed it to her teacher's aide, who reminded her that one of her fifth-grade pupils had developed a brain tumor and had died of it in 1986. At recess, she brought the diagram into the faculty room and placed it on a table. "All the teachers who saw it were shocked by what it showed," she remembers. "Almost everybody said that we should do something about it."

A week later, Buffo presented the concerns of the Slater teachers to the seven members of the Fresno Board of Education, who were holding their monthly meeting in the auditorium of the school-district headquarters, in the downtown section of the city. She followed Thom De-Young, of P. G. & E., who described the magnetic-field measurements that had been taken at the Lawless and Slater schools as relatively low compared to those emanating from household appliances. (DeYoung did not mention the fact that the magnetic fields emitted by most appliances fall off sharply within a few inches of their source.) "I began by telling the board members that an unusually large number of teachers at Slater had developed cancer, and that all of them had worked on the side of the school nearest the power lines," Buffo says. "Then I held up the diagram I had drawn and explained the significance of the red dots in Pods A and B and in the adjacent administrative areas. I went on to point out that quality reviews of schools in the district were almost always conducted by impartial outsiders, but that by asking P. G. & E. to do a survey of the power lines at Slater, school district officials had in effect allowed the utility to conduct its own quality review. When

I finished, the board members were staring aghast at the diagram, and there was a moment of stunned silence. Then they began to bombard me with questions, such as which teachers had fallen ill, what kinds of cancer they had developed, how far the power lines were from classrooms, and whether any children had been affected. I answered their questions to the best of my ability, but I declined to identify any of the afflicted teachers for reasons of privacy. At that point, they called DeYoung back to the podium and asked him if he had any explanation for what had happened. He told them that he did not, and that there was no definitive data that would allow anyone to determine a safe or unsafe level of exposure to the magnetic fields given off by power lines. The board members then directed Frank J. Abbott, the superintendent of schools, to investigate the situation and come up with a recommendation for how to deal with it."

On April 5th, school-district officials held a meeting at Slater for the purpose of devising a plan to deal with the power-line problem. It was chaired by Cathi Vogel, the chief financial officer of the district, and she was accompanied by an assistant school superintendent, William Hansen; Donald Beauregard, a school administrator; and Anna Phillips, a registered nurse, who was the director of the district's health services. George Marsh, Curtis Hurd, Patricia Berryman, and a teacher who had developed breast cancer after working in Pod A for eight years were on hand from the Slater School. Others in attendance were Gary Cardoza, the assistant director of the Fresno County Health Department; Dr. Hugh Stallworth, the Fresno County health officer; and Thom DeYoung.

At the beginning of the meeting, Vogel announced that it was closed to the public and the press, and proceeded to escort Amy Alexander, the *Bee*'s staff writer, from the room. During the discussion that followed, Phillips said that she had contacted Dr. Stallworth, and had learned that the county health department had a procedure for investigating and evaluating cancer clusters. Stallworth said that the county would gather data on individual cancer cases at Slater, and that if a cancer cluster was found to exist, he would contact officials of the state Department of Health Services and work closely with them to investigate it.

When DeYoung was asked to give P. G. & E.'s point of view, he said that the utility was unable to interpret the magnetic-field measurements it had taken at Slater, or to say whether there was any risk associated with any level of exposure. He said that the evidence about the magnetic-field hazard was inconclusive and contradictory, and that "our corporate stance is that the jury is out." Berryman pointed out that Slater was a year-round school, and that the utility had taken measurements there on only one occasion, in December. When she asked DeYoung if magnetic-field levels would not be higher in the summertime, he agreed that because of heavy air-conditioning use in the summer, this would undoubtedly be the case. Replying to a question about how the county health department intended to respond to the concerns of the Slater teachers, Stallworth said that the first order of business was to determine whether or not there was a cancer cluster at the school. He went on to say that just because some red dots on a map indicated a possible association did not necessarily mean that exposure to power-line magnetic fields had caused cancer among the teachers there. When asked whether the county's study could be expanded to include people living in neighborhoods near the power lines, Stallworth replied that this was possible. Cardoza, however, warned that such an investigation might cause panic in the surrounding community. Like Stallworth, he advised against making the assumption that power-line emissions could cause cancer.

By arrangement with the county health department, Anna Phillips came to Slater on the afternoon of April 10th to collect the names of teachers and students who were believed to have developed cancer, and to obtain authorization from them for their medical records to be released and reviewed. "She sat in Room 13 for about an hour after school had been let out, so that those of us who knew or had heard of colleagues or pupils with cancer could come by and provide her with information," Berryman recalls. "Right from the start, however, she refused to consider any name that was not accompanied by an address or a telephone number, claiming that there was no way for the school district to track down a teacher or student without it. As a result, we had early misgivings about how thorough the county's study would prove to be, but we kept silent about them because at that point we felt we should work within the system."

During the rest of April, Berryman and her colleagues waited to learn what measures the school district intended to take to protect them from the potential hazard posed by the power lines running past the Slater School. Meanwhile, fourteen teachers — most of whom worked in Pods A and B — asked to be transferred to other schools, and Katie Alexander, the first-grade teacher, died. Her funeral, which was held on April 27th, was attended by more than a dozen of her fellow teachers and scores of friends, who were described in an article that appeared in the Bee as wondering whether her death from brain cancer was connected with exposure to the magnetic fields given off by the power lines. Her son told the newspaper that school officials should close down the classrooms at Slater that were clos-est to the lines. "It's too late for my mother, but there are still so many others there who face the same danger," he said. Berryman remembers that Alexander's death had a profound effect upon her colleagues at Slater, because, as she has said, "We became all the more determined to remove ourselves and the children we were teaching from further exposure to the power lines."

CHAPTER ELEVEN

Safety for Our Kids

WHILE THE RESOLVE of the Slater teachers to do something about the power lines was stiffening, the seventeen members of the EPA's Scientific Advisory Board subcommittee who had been reviewing the E.P.A. draft report found themselves increasingly uneasy over its conclusion that power-line electromagnetic fields were a possible cause of cancer in humans. Following their initial meeting, in January, the subcommittee members had divided themselves into three panels: one of seven members, to assess the validity of the E.P.A.'s interpretation of the epidemiological evidence; another, also of seven members, to examine the Agency's interpretation of the experimental studies suggesting how electromagnetic fields might interact at the cellular level to cause or promote cancer; and the third, made up of the remaining three members, to determine whether the authors of the report had correctly evaluated the physics of how electromagnetic fields might affect biological systems. On April 12th and 13th, the full subcommittee met in San Antonio, Texas, to review the preliminary reports that had been written by each of the three panels.

The seven members of the epidemiology panel found that the authors of the E.P.A. draft had "achieved nearly complete coverage of pertinent work" but that their analysis of the epidemiological findings contained "too much unwarranted speculation about causal interpretation," which resulted in "giving emphasis to positive findings while de-emphasizing negative ones." This was a puzzling assertion, because the epidemiology panel failed to list any of the so-called negative findings. Nor did it acknowledge that five of the six childhood residential studies and a great majority of the thirty or so occupational studies that had been conducted since 1979 showed an association between exposure to power-frequency and other electromagnetic fields and the development of cancer. Indeed, no fewer than twelve childhood and occupational studies — all of them conducted, published, or re-analyzed between 1985 and 1989 — showed significantly increased rates of brain cancer among people exposed to electromagnetic fields at home and at work. Moreover, the steady accumulation of epidemiological evidence indicting power-line magnetic fields as a cause or promoter of cancer had been considerably strengthened since the middle of 1989 (the cut-off date of the E.P.A. report) by the studies showing an increase in breast cancer among men, and by a study that had been conducted by researchers at the University of Southern California School of Medicine. They found that men working for more than ten years as electricians, electrical engineers, and in other jobs involving prolonged exposure to strong electromagnetic fields were developing astrocytoma, a malignant tumor of the brain, at more than ten times the expected rate.

The report of the seven-member panel that assessed the experimental evidence relating to the interaction of electromagnetic fields at the cellular level also seemed inclined to extend the benefit of the doubt to electromagnetic fields. Its authors acknowledged that *in vivo* experiments had shown that power-frequency fields could produce fifty per cent decreases in melatonin levels in rats, and said these were "important biological effects." However, they went on to say that, because such findings did not "immediately implicate" electromagnetic fields in the development of cancer, the melatonin evidence should not have been presented as supporting evidence of carcinogenicity. This assessment of the melatonin findings seemed sanguine, considering the fact that a decrease in melatonin production had been strongly correlated with breast cancer both in rats and in human females, and in light of the recent male-breast-cancer data and the steep rise that had been occurring in the incidence of female breast cancer throughout the nation.

The review of the three-member panel on physics was notable for its insistence that special attention be paid to Robert Adair's claim that power-line magnetic fields below one hundred milligauss in strength could not produce biological effects, and for its suggestion that the credibility of the entire E.P.A. draft report "depends on whether this strongly held view by a distinguished physicist can be controverted or shown to be inapplicable." The conclusion that Adair's opinion should somehow negate the results of well over a hundred epidemiological and laboratory studies conducted by researchers around the world during a period of more than twenty years originated with Richard Wilson, the former chairman of the Department of Physics at Harvard University, who had been asked to serve on the S.A.B. review committee at the last minute, after physicists had complained that they were being deliberately excluded from its roster. In the end, Wilson's proposal, which would have given Adair's views a potential veto over the E.P.A. report, was deleted when some members of the subcommittee objected strenuously to it.

By then, however, it was clear that most of the seventeen subcommittee members had been persuaded that there was insufficient evidence to characterize electromagnetic fields as a possible carcinogen, and that the E.P.A. draft report should be made, as one of them put it, "a lot less inflammatory." Their decision to require a higher burden of proof before indicting the fields as cancer-producing should not have come as any great surprise, considering the intense and almost unprecedented opposition to the findings of the report which had been mounted by science and policy advisers of the Bush Administration, by the Air Force, and by many of the nation's physicists, as well as the notable lack of support the document had received from officials within the E.P.A. and other regulatory agencies. Further evidence of the hostility the report had provoked within the governmental-scientific community came when Matanoski revealed that some members of the S.A.B.'s Radiation Advisory Committee — the subcommittee's parent body, which is made up almost exclusively of scientists whose expertise lies in the field of nuclear and other ionizing radiation — were intensely opposed to its findings, and not only expected the subcommittee to issue recommendations for revising it, but also to provide justification for releasing it. Indeed, the pressures being brought to bear on the Scientific Advisory Board's subcommittee on electric and magnetic fields were thought to be such that in May, when the U.S. House of Representatives' Committee on Science, Space, and Technology authorized two million dollars for the E.P.A. to spend on electromagnetic field research for fiscal year 1992, it issued a report recalling the efforts of

D. Allan Bromley, President Bush's science adviser, to delay the release of the Agency's report and Crowell & Moring's stacking of witnesses at the January S.A.B. hearings, and informing Administrator Reilly that "this committee expects that the evaluation of scientific evidence and the transmission of expert consensus to the administrator by the [Scientific Advisory Board] will take place in a climate unaffected by political considerations." The committee went on to remind Reilly of his responsibility to shelter the Advisory Board "from undue outside influence."

Meanwhile, in Connecticut, new concern about the health hazard posed by power-line magnetic fields was raised when Maria Hileman, a staff writer for the *New London Day,* reported that three people living close to a high-current line that runs along Cliff Street in nearby Mystic, and supplies power for the town's business district, had died of primary brain cancer in recent years. Hileman went on to say that a preliminary survey showed that there had also been three prostate cancers, a breast cancer, and a death from renal cell carcinoma among residents of the neighborhood. When Hileman called the Department of Health Services to find out if it was going to investigate the situation in Mystic, Dr. Peter Galbraith told her that state health officials would use the Connecticut Tumor Registry to track cancer incidence and deaths in the town, but that they had decided not to conduct a statewide epidemiological study of the power-line hazard because many such studies were being conducted around the world. David Brown, chief of the Health Services' Division of Environmental Epidemiology and Occupational Health, told Hileman that the level of risk from exposure to power-line magnetic fields was "well below what one experiences from cigarette smoking." He did not say what cigarette smoking had to do with the development of the brain, breast, and prostate cancers that had afflicted people living near the high-current feeder line on Cliff Street, nor did he mention the epidemiological studies showing that the incidence of these three cancers was elevated in electric-utility linemen, telephone linemen, and other workers who were occupationally exposed to power-line magnetic fields.

On the other side of the continent, the teachers at the Slater School were growing impatient with the delay of Fresno Unified School District officials in devising measures to protect them. The situation reached a flash point on the afternoon of May 7th, 1991, when Cathi Vogel and William Hansen met with George Marsh and the members of the Slater faculty in one of the bungalows that house classrooms on the northeast side of the school. "Vogel seemed taken aback when some of us told her at the outset that we didn't feel safe working in Pods A and B," one of the teachers recalls. "She was even more taken aback when Charleene Conley, whose eight- and ten-year-old sons attended the school, asked what the district intended to do to insure the safety of the children there. Vogel maintained that moving classrooms from Pods A and B would be prohibitively expensive, and that the district could do nothing about the power-line problem in the near future. When someone suggested that perhaps we should take our concerns back to the Board of Education, she became indignant, and told us that we had no right to do so without her permission."

After the meeting, the Slater teachers gathered in their faculty room and decided to put the school district on notice that they meant business. They drew up a memorandum declaring that as of August 10, 1991, "We, the faculty of Slater Elementary School, will not teach in Pods A & B (rooms 1–10)." The memorandum was signed by all the members of the faculty who were on duty at the time — forty-seven teachers and teachers' aides — and it was sent to Olivia Palacio, the superintendent of the "school pyramid" to which Slater belonged (there were then four such pyramids in the Fresno Unified School District), with carbon copies to Vogel, Hansen, Beauregard, and the members of the Board of Education.

Meanwhile, Charleene Conley had gone home from the May 5th meeting and called the Board of Education to request that Slater parents be placed on the agenda to speak at the next Board meeting, which was scheduled for May 23rd. When the scheduling secretary said that this would not be possible, Conley threatened to set up a picket line in front of school-district headquarters and to inform local television stations that Slater parents were being prevented from speaking. At that point, the secretary relented and asked Conley to send her the names of any parents who wished to address the Board.

On May 9th, Conley arranged to have a hand-written notice sent home with each child who attended the Slater School. The notice informed parents that the first meeting of an organization called Parents for a Safe Environment would be held in the school's all-purpose room at 7 P.M. on the following day. The meeting was attended by more than seventy-five parents, who agreed that the south side of Slater should be shut down until studies of the power-line hazard were completed. A few days later, seventy-eight parents signed a petition telling the Fresno Board of Education that they did not want their children to attend class or play on the south side of the school.

On May 14th, a hundred parents and children held a rally at the school. After cordoning off the south side of the campus with yellow tape and hanging a huge sign that read "DANGER, ELECTROMAGNETIC FIELD" on the chain-link fence on Emerson Avenue, they marched up and down the sidewalk carrying signs that warned of the power-line health hazard, and chanting "Safety for our kids!" Just before the rally began, school administrator Donald Beauregard arrived on the scene and told half a dozen television and newspaper reporters who had come there to cover the rally that the school district had decided to ask the Board of Education for permission to close the ten classrooms on the south side of the school, and to place portable classrooms to accommodate displaced students at the other end of the campus. Beauregard said that the district would also recommend closing the playground near Emerson Avenue and moving its equipment elsewhere. He went on to say that district officials were moving ahead with their plan even though there was no conclusive proof that power lines could cause cancer. The headline of the story that appeared in the *Bee* the next day read, "School district bows to power-line fear."

On Sunday, May 19th, the *Bee* ran a front-page story about the power-line hazard by a reporter named Russell Clemings, who started out by reminding his readers of the school district's decision to close the ten classrooms at Slater. Clemings went on to say that although epidemiological studies had shown that children and workers exposed to power-line magnetic fields had an increased risk of developing cancer, most scientists believed that the research was still "too rudimentary" for them to reach a consensus on the extent of the problem. He then quoted Dr. Neutra, of the California Department of Health Services, whom he described as a scientist who advises public officials about the power-line hazard, and who considers his job to be politically perilous. "I guess my worst nightmare is that we decide to do something that's very expensive, and then we discover that it was the wrong thing to do and we made things worse," Neutra said. He added, "The contrary mistake is not doing enough, and we worry about that too."

On Monday, May 20th, Dr. Neutra and Dr. Eva Glazer, a medical epidemiologist with the Department of Health Services' Cancer Surveillance Section, came to the Slater School at the invitation of Dr. Stallworth, of the county health department, to attend a meeting of representatives from the school district, the Slater teachers, the Slater Parent-Teacher Association, Parents for a Safe Environment, and P. G. & E., who had gathered there to receive a report on how the cancer-cluster investigation had progressed since its start in April. As in the case of earlier meetings, members of the press were barred from attending by Cathi Vogel, and she also advised the participants against speaking to the media, on the ground that premature disclosures about the cancer study might cause unwarranted fear in the surrounding community. Once the meeting got under way, Stallworth said that there did not appear to be a cancer cluster at Slater, because no single type of cancer had been found to predominate among the teachers and employees there. This assessment was supported by Dr. Glazer and Dr. Neutra, and they also said that the county health department investigation, like most studies of the power-line hazard, would probably turn out to be inconclusive.

When someone asked Dr. Glazer, who was sitting directly in front of a chalkboard holding Doris Buffo's diagram, what explanation she had for the fact that all of the cancer among Slater teachers and employees had occurred among women working on the side of the school next to the power lines, she replied that this was something that re-

mained to be determined. When Berryman pointed out that
she and several teachers who had worked in Pods A and B
had developed nonmalignant tumors of the breast, and that
several other teachers and administrative personnel who
had worked on the south side of the school had developed
nonmalignant tumors of the uterus, Glazer replied that epidemiological guidelines for the evaluation of a cancer cluster required that only malignant tumors be included in the
study.

When Berryman again raised the question of whether
the magnetic-field levels recorded by P. G. & E. the previous
December accurately reflected the levels that existed at the
school throughout the year, Neutra suggested that Slater
teachers take their own readings and keep a record of
them. (After the meeting, he loaned Berryman a gaussmeter and showed her how to use it.) He also proposed that
a subcommittee made up of people at the present meeting
be formed to review the situation at Slater on a periodic
basis, and he accepted Charleene Conley's invitation to sit
on it as a representative of the parents. (Neutra's proposal
for the establishment of a subcommittee was in keeping
with a strategy he had developed earlier in his career for
dealing with California citizens who were suspicious of the
efforts of the Department of Health Services to study diseases associated with chemical waste disposal sites. In an
article entitled "Epidemiology for and with a Distrustful
Community," which appeared in *Environmental Health
Perspectives* in 1985, he pointed out the importance of taking positive steps "to alleviate the antagonism and to involve the community in an active and constructive role in
the epidemiologic study." Among the steps he recommended were the formation of "a citizen and industry
advisory committee" that would hold regular meetings in

order "to involve community leaders, including those distrustful," and to "conduct an epidemiological census and a neighborhood environmental exposure survey." According to Neutra, such a strategy would produce "a defusing of the antagonism toward authorities," and result in a "complete acceptance by the community of the merit of the report" that the Department of Health Services would issue on the study it had conducted.) Conley and Berryman, who were unfamiliar with any of Neutra's previous pronouncements about the power-line health hazard, were heartened by his willingness to involve himself and the Department of Health Services in the situation at Slater. "We knew that Dr. Neutra was a senior state health official, and we welcomed his interest as an indication that our concerns would finally be addressed," Berryman recalls.

The next meeting of the Board of Education, which was held on May 23rd, was attended by a standing-room-only audience that included nearly a hundred Slater parents. At the outset, Dr. Stallworth said that he and his staff had confirmed seven cases of cancer among Slater teachers and one case of cancer in a Slater student, but that this number of cases did not constitute a cluster, which he defined as "an unusual occurrence of cancers." He went on to say that his conclusion was supported by Dr. Neutra, whom he described as "probably the most knowledgeable person in California on electromagnetic fields." Stallworth also said that there were no definitive data to suggest that power-line magnetic fields had a negative effect upon human health. When Juan Arambula, the president of the school board, asked him to submit a written report of his findings, he said that he had not considered a report to be necessary, because it was obvious to him that there was no cluster. Arambula then asked Stallworth why he felt documentation was unnecessary when so many teachers at Slater had fallen ill, and Stallworth replied that the health department was still reviewing the situation, and that a report would be forthcoming. (Two months later, one of his assistants said that the documentation for the report had been lost in Stallworth's computer.)

When another member of the school board asked Stallworth how he had arrived at the conclusion that there was no cancer cluster at Slater, Stallworth explained that the teachers there had not been afflicted with a predominance of any single type of cancer. When someone from the audience asked Stallworth if he could explain why all the cancers that had been reported had occurred among women who worked on the side of the school nearest the power lines, there was prolonged applause, as it proved to be a question for which Stallworth had no answer. Shortly thereafter, Superintendent Abbott declared that because of the concerns of the Slater parents and disruption to the educational program at the Slater School, he believed it important to move students and teachers out of the classrooms near the power lines until more definitive information was available. The members of the Board of Education then voted unanimously to adopt the resolutions that had been put forth by the school district to accomplish this.

CHAPTER TWELVE

All Your Ducks in a Row

THE FIRST MEETING of the Slater Electromagnetic Field Study Subcommittee — the group that had been formed at Neutra's suggestion — was held at the school on June 5th. It was presided over by Donald Beauregard, and was attended by Neutra, Berryman, DeYoung, and representatives of the county health department, Parents for a Safe Environment, the Slater teachers, the teachers' union, and the Slater PTA. Beauregard began the proceedings by announcing that Neutra had sent some research information to the school district that would be copied and distributed to all subcommittee members within a week. (The information turned out to include a proposal by the Department of Health Services that the California State Legislature finance a two-hundred-thousand-dollar pilot study in which children and teachers in selected schools — some near power lines and some not — would wear gaussmeters for twenty-four-hour periods, in order to determine whether the average magnetic-field exposure of children in schools close to high-current power lines was different from that of children who did not attend such schools, and whether a series of inexpensive spot measurements could be used to estimate the accumulated exposure of a child to magnetic fields over the course of a day, thus simplifying a large-scale statewide study of the power-line hazard in schools, should one become necessary.) Beauregard then announced that five temporary trailers, each containing two classrooms, would be set up on the Slater campus as soon as possible to house the teachers and students who were scheduled to be moved from Pods A and B. Berryman reported that she had found magnetic-field levels at the proposed site of the trailers to be considerably lower than those in her classroom in Pod A, where, using the gaussmeter that Neutra had lent her, she had measured fields of between one and a half and three milligauss on seven consecutive days within the previous two weeks. Neutra said that the electromagnetic fields at Slater required further study, and he suggested to DeYoung that P. G. & E. place meters for extended periods of time at various locations on the school campus, to record magnetic-field levels that were being given off by the nearby power lines. DeYoung and his associates at the utility did not get around to taking these measurements for nearly three months. Within two weeks, however, P. G. & E. had mailed the pamphlet that Neutra and his colleagues in the Department of Health Services had written a year earlier, telling Californians that it was not possible to characterize any given magnetic-field levels as either safe or dangerous, to eighty-six hundred Fresno homes.

At the invitation of Parents for a Safe Environment, Duane A. Dahlberg, an associate professor of physics at Concordia College, in Moorhead, Minnesota, who had been studying the effects of electromagnetic fields on humans and farm animals for a number of years, took magnetic-field

readings at the Slater School on June 27th and 28th. Dahlberg measured fields of between one and 1.7 milligauss in
Pods A and B, and fields of between 0.3 and 0.6 milligauss
in Pods C and D. In Pod D, he found a number of significant
"hot spots" — localized areas where strong magnetic fields
were being given off, usually as a result of imbalanced currents in electrical wiring running through the floors and
walls. In subsequent reports, Dahlberg said that the ambient magnetic fields he had measured in Pods A and B
were associated with low to medium power in the nearby
transmission lines, and that at times of increased power use
the sixty-hertz magnetic fields given off by the lines would
be "proportionally higher." He noted that the perception of
teachers and parents was that "children in Pods A and B had
more health problems, and did not learn as easily, and had
different behaviors than in Pods C and D." He also stated
that in his opinion there was a possibility that the transmission lines were implicated in the cancer cluster among employees on the south side of the school.

Meanwhile, Neutra had telephoned Berryman on several
occasions to find out what magnetic-field levels she was re-
cording in her classroom with the gaussmeter he had
loaned her. Berryman told him that she was continuing to
record levels of between one and a half and three milligauss throughout the room. Since she had never seen the
EPA report, she did not know that these levels were roughly
equal to those associated with an increased risk of cancer
among children living in homes close to high-current
power lines. Nor was she aware that they were about equal
to the average magnetic-field levels that had been measured
in high-current homes in Denver, where women were
found to be suffering from excess cancer of the breast and
uterus.

Berryman also informed Neutra that on July 5th — a hot day in Fresno, when air-conditioning use was heavy — the first-grade teacher with whom she had left his meter when she had gone on vacation had measured magnetic-field levels of between three and a half and four milligauss in an adjacent classroom. Hot days occur frequently in Fresno between May and October, and four milligauss is almost equal to the average daily exposure levels of the forty-five hundred New York Telephone Company cable splicers in whom Matanoski and her colleagues at Johns Hopkins had found the incidence of leukemia to be seven times higher than expected, and cancer of many other types to be elevated.

At no time did Neutra, who had by now taken over the direction of the cancer-cluster study at Slater, express any concern to Berryman over the magnetic-field levels she had measured, or inform her that they were approximately the same as levels that had already been associated with a significantly increased cancer rate in children. Nor did he tell Berryman that during 1989 and 1990 he and some colleagues in the Department of Health Services' Environmental Epidemiology and Toxicology section had conducted an eighteen-month-long investigation of a cancer cluster at the Montecito Union School — the elementary school in Montecito, near Santa Barbara — where magnetic-field levels in several classrooms on the north side of the school were slightly lower than those measured in the Pod A classrooms at Slater by Berryman. Between 1981 and 1988, six cases of cancer were known to have developed among children who had attended Montecito Union, which is situated close to a substation and within forty feet of a sixty-six-thousand-volt feeder line carrying high current from a nearby substation. Two children developed leukemia; three children developed lymphoma; and one child developed testicular cancer. (A teacher's aide with several years of experience in the kindergarten subsequently developed a brain tumor.) According to statistics compiled by the National Cancer Institute, the incidence of cancer among the children at the school was more than fifteen times the expected rate. In June of 1990, Neutra appeared at a meeting of the school board and the parents' task force, and insisted that not enough was known about the effects of electromagnetic radiation to say if the magnetic-field levels at Montecito Union posed a health hazard. Neutra said that it would take at least two more years before enough data were collected to make such a judgment. In December, a few days before Berryman and her fellow teachers at Slater read about the EPA report and first began to suspect that emissions from the high-voltage transmission lines near Pods A and B might account for the high incidence of cancer among the women who worked there, Neutra and his colleagues issued the final report of the investigation they had conducted at Montecito Union, in which they said that because of "uncertainty" about whether there is a hazard from magnetic fields, they were not prepared to make recommendations about safe levels of exposure.

Some observers believe that the neutrality maintained by Neutra and his associates toward the power-line hazard was dictated by higher-ranking officials of the Department of Health Services and the California Public Utilities Commission, who, fearful of arousing public concern, did not want anything said that might be construed as an admission that power-line magnetic fields could pose a health risk. Be that as it may, public concern over the hazard had become so

pronounced that on January 15, 1991, the Public Utilities
Commission announced that it was opening an investigation "to explore the scientific evidence relating to possible
health effects, if any, of utility employees' and consumers'
exposure to electric and magnetic fields created by electric
utility power systems," and to examine "the range of regulatory responses which might be appropriate." A press
release that accompanied the announcement said that the
Department of Health Services and the Public Utilities
Commission were jointly conducting three research studies
of the health risks associated with exposure to power-line
magnetic fields, and that depending on the results of these
and other studies, the Commission would determine
whether it should change its existing rules, regulations, and
policies regarding the design, siting, and operation of electric power facilities.

As it happened, the three studies were being coordinated by Neutra, and on March 29th he responded to the
Commission's request for comments from interested parties by writing to Steven A. Weissman, the administrative
law judge who had been placed in charge of the Commission's investigation. In his letter, Neutra proposed that the
Commission "convene a standing group of 'stakeholder'
advisers from industry, labor, environmental groups, public
service organizations, concerned citizens, rate payers, the
media, and politicians from both parties," who would plan
the process by which the power-line health issue could be
investigated in greater depth. He went on to describe this
process as one in which the stakeholder-advisers would try
to achieve consensus between the opposing views of utility
officials and public-sector advocates on such issues as
whether there was a power-line hazard, what options were
available to lessen the magnetic-field exposure of workers

and consumers, and what the economic consequences of such mitigation might be. "We feel that California has the opportunity to see that a comprehensive examination of options is carried out, debated, and understood by the major stakeholders over the next year and a half, so that if hazard and mechanism research warrants it, we will be in a position to take well-informed action," Neutra wrote. He went on to assure Judge Weissman that the California Department of Health Services was prepared to assist the PUC investigation by overseeing the evaluation of the health hazards associated with power-frequency magnetic fields, and the biological mechanisms through which they might be taking place.

Two weeks earlier, Neutra had described some of the difficulties he faced in dealing with the power-line health issue at a conference on electromagnetic fields that was co-sponsored by the Los Angeles Department of Water and Power and the California State University's Statewide Energy Consortium, and held at the Water and Power Department's headquarters building, in Los Angeles. "There are kinds of environmental issues out there that you could spend your time on, and if you choose the wrong one you may miss something else on which you might have better spent your time," Neutra told his audience. "You also realize that ultimately this becomes a regulatory thing, and if you go to bat too early, before you have all your ducks in a row, you may find yourself spending five years and losing a battle that would be easier to win when there's more evidence." During a discussion period, Neutra expressed his dilemma in more personal terms. "I, for example, have an Epson printer that gives off a pretty good sixty-hertz field, and I moved it to a shelf that's farther away from where I sit," he said. "I explain to people why I did this, but then there is some possibility that maybe I made the wrong decision on that. It turns out in my new office that the fluorescent lights give me a two-and-a-half-milligauss field. I haven't turned off the lights in my office and I haven't demanded a new office. It's just too much of a hassle. So, you could say that that's not completely rational, but there's a kind of rough cost-benefit behavior there."

Four months later, Neutra's personal and professional ambivalence toward the power-line health issue remained unknown to Berryman and her colleagues at the Slater Elementary School, who welcomed his decision to take over the cancer-cluster study as indicating that their concerns would be thoroughly investigated by the Department of Health Services. At first, their optimism appeared to be justified. On July 24th, Neutra and Glazer called Berryman, who was at home on vacation, to find out whether her list of reported cancer victims at Slater tallied with the list that had been sent to them by Anna Phillips. The two lists proved to be identical; a total of seventeen teachers, teacher's aides, and administrative staff members, as well as twelve former students, had by then been reported to the state officials as possibly having developed cancer. All of the teachers, teacher's aides, and staff members had worked on the side of the school nearest the power lines. When Neutra asked Berryman where a nurse's aide who had developed cancer had worked, she told him that the aide's office had been in the administrative area, about ten to fifteen feet from Pod A. After discussing the location of the office used by two administrative secretaries, who had also been reported to have developed cancer, Neutra asked Berryman how many people had worked in the administrative area since 1972, when the school had opened. When Berryman said she didn't know, Neutra asked her if she thought six

would be a reasonably accurate estimate, and she agreed
that it might be.

A few days later, Neutra called Berryman to ask if she
could put together a list of all teachers and teacher's aides
who had worked at Slater since 1972. He said that he
needed such a list in order to complete the cancer study,
but that school-district officials had claimed that compiling
it would require them to conduct a manual search of the
district's entire personnel files — an undertaking they considered too expensive and time-consuming. Berryman, in-
terrupting her vacation, enlisted the help of Loretta Hutton,
a second-grade teacher who had been at Slater for seventeen years, and, together with some colleagues, they spent
several afternoons during the first week of August combing
through school yearbooks that were stored in the main administrative office.

"Starting with the 1972–73 yearbook, we identified
ninety-five teachers and forty-four teacher's aides whose
photographs appeared in class pictures taken through the
1990–91 school year," Berryman remembers. "We also
counted the number of students who appeared in the class
pictures. We wrote the names of the teachers and teacher's
aides and the number of children in each of their classes on
three-by-five cards. That evening, I took the cards home and
sorted them to find out how many years each teacher and
teacher's aide had worked at Slater, and who among them
had worked in Pods A or B, and for how long. The next day,
two of my fellow teachers fed the names into a computer
program that arranged them in alphabetical order, and
printed out a master list, which we gave to Anna Phillips,
who forwarded it to Dr. Neutra on August 7th. At about this
time, Dr. Neutra sent me a standard illness-report form,
which I photocopied and distributed among forty or so of

my fellow teachers, so that they could list any health problems they might have encountered. The following week, Dr. Glazer called to request a year-by-year count of the number of children who had attended the school, together with a year-by-year count of incoming kindergartners and outgoing sixth graders, which she needed to estimate the rate of transiency among the student population. These figures were sent to her on August 22nd. That same day, Anna Phillips asked me if I could help her trace the whereabouts of half a dozen teachers and children, who had been reported as having developed cancer, but for whom she had no addresses or telephone numbers. As it turned out, I didn't have any information that would be of help, but her request got me to wondering why, at that late date, state officials had not become involved in tracking these people down, and whether they were making any effort to do so. I knew that unless these teachers and students were contacted, their cases verified, and signed authorizations obtained from them or their guardians, they would not be included in Dr. Neutra's study. I was also aware that at least one Slater staff member — someone we knew for certain had developed cervical cancer — had refused to sign an authorization form for reasons of privacy. In addition, I realized that since the study was dealing with the occurrence of cancer in a relatively small number of people — about a hundred and fifty teachers, teacher's aides, and staff members — the exclusion of any cancer case, for whatever reason, was bound to affect its outcome. So once again I began to wonder just how complete and valid the investigation we had worked so long and so hard all these months to bring about would prove to be."

CHAPTER THIRTEEN

Traversing the Tightrope

A FTER A THIRD, and final, meeting that was held in Arlington, Virginia, from July 23rd to 25th, the seventeen members of the Scientific Advisory Board subcommittee concluded that the E.P.A. draft report on the carcinogenicity of electromagnetic fields had "serious deficiencies," and that there was "insufficient evidence from the human epidemiologic data and from animal/cell experiments to establish unequivocal cause-and-effect relationships between low-frequency electric and magnetic field exposure and human health effects and cancer." The fact is that the authors of the E.P.A. report had never claimed that power-line electromagnetic fields were an "unequivocal" cause of cancer, but only a "possible" cause of cancer, and the subcommittee had not been asked to determine whether the evidence of carcinogenicity was "unequivocal" but simply to review "the accuracy and completeness of the entire document and . . . whether the interpretation of the available information reflects current scientific opinion."

The subcommittee members also said that although some epidemiological studies reported an association be-

tween living close to power lines and "an increased incidence of some types of cancer," the E.P.A. report's
conclusion that power-line magnetic fields were a possible
cause of cancer was "currently inappropriate because of
limited evidence of an exposure-response relationship and
the lack of a clear understanding of biologic plausibility." At
the same time, they covered all bets by warning that because no factors other than electromagnetic fields had been
identified to explain the excess cancer risks found in the
studies, the existing evidence that exposure to the fields
was associated with the development of cancer "cannot be
dismissed."

The question of who was going to take the lead in determining the extent of the public health hazard posed by
power lines and other sources of electromagnetic-field exposure had been bandied about for months in the Congress. Some members wanted the E.P.A. to direct the effort,
claiming that in terms of its mission and mandate the
Agency was the logical choice for the job. However, the De-
velopmental and Cell Toxicology Division of the E.P.A.'s
Health Effects Research Laboratory, at Research Triangle
Park, in North Carolina, which had once housed the largest
and best electromagnetic research group in the world, had
been virtually shut down since 1985, as a result of budget
cuts imposed by the Reagan Administration, and high
Agency officials had been so traumatized by the Bush
Administration's attempt to suppress the draft report on the
potential carcinogenicity of electromagnetic fields that they
clearly wanted no part of a program that could be expected
to provide major political and economic repercussions. Indeed, as Louis Slesin pointed out in an editorial entitled
"EMF Research: Who's in Charge?" which appeared in the
July/August, 1991, issue of *Microwave News,* "Today, a year

after EPA released a draft report linking EMFs to cancer, the agency is still not supporting a single EMF experiment in its own labs."

Other members of Congress were in favor of farming out the electromagnetic research effort to some nongovernmental organization. To this end, the Large Public Power Council — an organization that represents seventeen of the largest public utilities in the nation — had been hard at work during 1990 to set up a national EMF research program that would be jointly financed by the federal government and the utility industry. Late in that year, Council officials had approached the Health Effects Institute (HEI) — a nonprofit research outfit located in Cambridge, Massachusetts, whose four-member board is chaired by Archibald Cox — to see if the Institute would be interested in coordinating such a program. Early in 1991, HEI began to study the feasibility of the project, with the help of a grant of more than half a million dollars, which was supplied by the E.P.A. out of a seven-hundred-and-fifty-thousand-dollar appropriation that the Congress had given the Agency for EMF research, and with matching funds from the utility industry.

The Health Effects Institute had been established in 1980, in order to carry out its self-described mission of ensuring that "objective, credible, high-quality scientific studies are conducted on the potential human health effects of motor vehicle emissions," and by so doing to "ease the adversarial atmosphere that existed between the manufacturers of motor vehicles and the Environmental Protection Agency." The Institute's efforts in this regard were financed equally by the E.P.A. and by more than two dozen automotive manufacturers and marketeers, and they were evidently considered to be satisfactory, because, in 1989, the

E.P.A. entrusted HEI with the mission of conducting an objective study of the health hazards associated with the inhalation of asbestos fibers present in buildings across the nation. The Institute's asbestos-research program, known as HEI-AR, was financed by two million dollars in public funds from the E.P.A., and by matching funds from undisclosed private sources.

During the winter and spring of 1991, officials of the Health Effects Institute conducted an aggressive campaign aimed at securing a contract to direct the national electro-magnetic-field research program, and the Institute soon became the leading candidate for the job. As part of the campaign, HEI issued a preliminary research plan which declared that the Institute's matched public/private financing would guarantee the independence of its research into the biological effects of electromagnetic fields. HEI's reputation for independent research was soon compromised, however, by disclosures that were made in an article about its asbestos program, which appeared in the January 1991 issue of *Asbestos Issues*. The article was written by Edward J. Westbrook, an attorney with the Charleston, South Carolina, law firm of Ness, Motley, Loadholt, Richardson & Poole, who said that a Freedom-of-Information request to the E.P.A. had produced an HEI document revealing that ninety-five per cent of the private financing for the Institute's asbestos research was coming in equal shares from asbestos manufacturers, insurance companies, and real estate interests, which had a vested interest in minimizing the health hazard posed by asbestos in buildings. Westbrook went on to say that seven of the fourteen active members of the HEI-AR Asbestos Literature Review Panel had been listed as expert witnesses for the asbestos industry in the Philadelphia Class Action — a combined lawsuit brought against the industry by fourteen thousand school districts across the nation, seeking damages for the cost of finding and removing asbestos-containing materials from school buildings — and that since being appointed, several members of the HEI panel had testified in behalf of asbestos manufacturers who were defendants in property-damage lawsuits brought by building owners.

Westbrook called for HEI officials to remove all panelists who were connected with asbestos litigation, and he suggested that if this were not done the Institute should be investigated by the Congress. A few months later, the National Association of Attorneys General — a group whose members were bringing lawsuits against asbestos manufacturers whose products had been found in government-owned buildings — adopted a resolution expressing "deep concern" that HEI-AR was "incapable of producing an objective, unbiased report" on the hazard of asbestos in buildings, and calling upon the E.P.A. and Congress "to disqualify from serving on the HEI-AR Literature Review Panel any expert who has testified or has agreed to testify for either the plaintiff or the defense in asbestos-related litigation."

Because of allegations that its asbestos-research program was tainted by pro-industry bias, HEI officials were subjected to intensive questioning by the members of a steering committee, who met in Washington, D.C., on June 27th and 28th to set up a twenty-five-million-dollar National Electromagnetic Field Research Program. The program was being organized through the joint efforts of the Large Public Power Council and the New York State Department of Public Service, whose staff members expressed concern about HEI's ability to direct open and objective EMF research. Not surprisingly, support for HEI's bid to direct the national program dwindled during the summer, and by late

July, when the Scientific Advisory Board's subcommittee on electric and magnetic fields met to conclude its review of the E.P.A.'s draft report, several of its members expressed serious reservations about the Institute's proposed role in the national program.

Exception to the criticism of the Health Effects Institute's participation in the program was taken by M. Granger Morgan, a professor of electrical and computer engineering, and the head of the Department of Engineering and Public Policy at Carnegie Mellon University, in Pittsburgh. Morgan had been studying the electromagnetic-field problem since 1982, when he had been asked by the Department of Energy to perform risk analysis of the potential health hazards posed by exposure to power-line emissions. He argued in favor of an HEI role in the national program, on the ground that the Institute would provide a "buffer" between the private sector and the E.P.A., and thus enable the Agency to concentrate on long-term EMF research. Whatever the merit of his claim, the fact was that the Institute had become a leading candidate to win a contract for coordinating the multi-million-dollar proposed national program, and was thus an important potential source of research money. Morgan's situation was similar to that of many other EMF researchers, who were continually faced with the prospect of achieving a delicate balance between when and how to speak out about the problem they were studying, and the need to acquire financing for their investigations from industry, government, and institutional officials, who often seemed to be more fearful of the political and economic consequences of alarming the public than of the possibility that power-line fields might pose a major public health hazard. Over the years, however, his attempts to maintain equilibrium while traversing the tightrope of the electromagnetic-field controversy had been revealing.

In May of 1986, Morgan wrote an editorial for *Science,* the official publication of the American Association for the Advancement of Science, in which he declared that government agencies should henceforth consider "stopping rules" for terminating risk research, on the ground that such research "may sometimes do more harm than good." In the editorial, Morgan claimed that several years of expensive research into the question of whether electromagnetic fields posed a health hazard had produced ambiguous or "decidedly inconclusive" results that had, nonetheless, alarmed the public and encouraged litigation to be brought against the nation's utilities. In March of 1988, however, he and several of his colleagues at Carnegie Mellon wrote an article for *Public Utilities Fortnightly,* in which they declared that "the results from the epidemiological studies are grounds for concern," especially in light of experimental research showing that electromagnetic fields could cause biological effects at the cellular level. They went on to suggest that the number of people exposed to transmission-line fields could be reduced by charging utilities an exposure fee for every person living within a certain distance of a proposed new line.

Up to this point, Morgan's work had been financed mostly by the National Science Foundation and the Department of Energy. However, during 1988, Morgan's Department of Engineering and Public Policy obtained a grant of more than $300,000 from the Electric Power Research Institute, as well as a contract with the Office of Technology Assessment (OTA) of the United States Congress. The OTA contract resulted in a one-hundred-and-three-page report entitled "Biological Effects of Power Frequency Electric and Magnetic Fields," which was made public in June of 1989.

In the report, Morgan and two coauthors said that a "growing number of positive findings have now clearly demonstrated that even weak low-frequency electromagnetic fields can produce substantial changes at the cellular level," and that "the emerging evidence no longer allows one to categorically assert that there are no risks." They went on to describe a "prudent avoidance" strategy, declaring that "by avoidance we mean taking steps to keep people out of fields, both by re-routing facilities and by redesigning electrical systems and appliances," and that "by prudence we mean undertaking only those avoidance activities which carry modest costs."

Meanwhile, Morgan had used money from the National Science Foundation and EPRI to write a forty-five-page brochure entitled "Electric and Magnetic Fields from 60 Hertz Electric Power: What do we know about possible health risks?" This brochure was published in March of 1989 and it was widely distributed by EPRI and its utility-industry sponsors. (It was from this brochure that David Brown and his colleagues in the Connecticut Department of Health Services borrowed the analogy between the advent of sunrise at the rooster's crowing and the development of cancer following exposure to power-line magnetic fields.) In it Morgan defined his concept of prudent avoidance in a manner that was considerably more palatable to the power companies than the definition that appeared in the OTA report. "If you are buying a new home it might be prudent to consider the location of distribution and transmission lines as one of the *many* things you consider," he wrote. "However, remember that even if fields are ultimately demonstrated to pose a health risk, things like traffic patterns in the streets and radon levels in the house are likely to be more important for your own or your children's overall safety than anything related to fields. If you are already in a home, moving in order to get away from existing lines goes beyond what we would consider prudent."

At an EPRI-sponsored seminar for utility officials, which was held in Delavan, Wisconsin, on September 28th, 1989, Morgan described a series of strategy options that power companies should consider in response to the health risks posed by power-line emissions. The options ranged from outright denial ("try to ignore the issue and hope that it will go away"), to modified denial ("acknowledge the reality of people's fears and try to persuade them these fears are groundless"), to calling for more research ("a very tangible way in which a utility can demonstrate concern and a commitment to action"), to seeking the redesign of household wiring and some electrical appliances (on the grounds that "power lines are not people's only source of exposure"), to undertaking to reduce exposure by rerouting, reconfiguring, or, if necessary, burying transmission and distribution lines. Regarding the latter prospect, Morgan acknowledged that some members of his audience might feel that "even suggesting such options verges on the irresponsible, on the grounds that talking about them may increase the chances that such options will actually become necessary." He went on to say that the value of electric power to society "dramatically outweighs even the worst conceivable health consequences of field exposure," but that if power-line emissions should prove to be a public health hazard, this fact "will not mean that we can continue to do business as usual."

In August of 1990, Morgan and Indira Nair, a physicist who is the associate director of Carnegie Mellon's Department of Engineering and Public Policy, wrote a nine-page report entitled "Electromagnetic Fields: The Jury's Still

Out," which appeared in *IEEE Spectrum,* a monthly publication of the Institute of Electrical and Electronics Engineers, Inc. In their article, Morgan and Nair listed or described more than two dozen epidemiological studies that showed an association between residential and occupational exposure to power-line electromagnetic fields and the development of leukemia, brain cancer, and other malignancies, and said that most of the experts they had talked with "privately give odds somewhere between 10 and 60 percent that within the next decade it will become clear that exposure to fields produces significant health risks." The two researchers stressed the importance of making prudent avoidance feasible, so that utilities could take responsible actions now "without fear of adverse legal or regulatory consequences." Among the prudent-avoidance steps they recommended for consideration were locating power lines away from areas where people live, requiring wider rights-of-way for transmission lines, and burying transmission and distribution lines in areas of high-density population.

In November of 1990, Morgan's thoughts about prudent avoidance were described somewhat differently in the same San Diego Gas & Electric brochure that quoted Neutra as being opposed to moving power lines. According to the utility brochure, Morgan advised people to unplug their electric blankets, move motor-driven clock radios from their bedside tables, and keep farther away from the display monitors of their computers. The brochure made no mention of what avoidance measures Morgan advised people to take with regard to power lines, but it described him as believing that "even under the most pessimistic assumptions, it's hard to justify the costs of moving power lines or modifying old facilities."

Considering the contradictory nature of Morgan's

assessments of the power-line hazard over the years, it seems ironical but not surprising that in the summer of 1991 he should join his colleagues on the Scientific Advisory Board's subcommittee in telling the E.P.A. that its report on the potential carcinogenicity of electromagnetic fields required "logical reorganization and complete rewriting with particular attention to careful and precise use of language." In arriving at the conclusion that power-line fields were a possible cause of cancer in humans, the authors of the E.P.A. report had pointed out that the results of the childhood-cancer studies were supported by the findings of occupational studies showing that workers exposed to power-line and power-frequency fields were experiencing significantly elevated risks of developing leukemia and brain cancer, as compared to workers in nonelectrical occupations. Curiously, however, Morgan and his colleagues on the Scientific Advisory Board subcommittee made no mention of the occupational studies in their review of the E.P.A. report — indeed, one member of the subcommittee had said that the subject of the occupational studies was scarcely raised during the group's discussions — so the strong correlation between the findings of the childhood residential studies and the adult occupational investigations was totally ignored.

As for animal studies, the subcommittee members pointed out (as the authors of the E.P.A. report had done) that no experiments had ever been conducted to determine whether lifetime exposure to magnetic fields could result in the development of excess cancer in test animals. The fact that this should be the case, twelve years after the publication of the first epidemiological survey linking residential exposure to power-line magnetic fields with an excess risk of leukemia and other cancer in children, reflected an appalling lapse on the part of the utility industry and EPRI, as well as on the part of the E.P.A. and other governmental agencies, which should have undertaken to conduct animal studies of the carcinogenicity of power-line emissions at least by 1987, when the findings of the first childhood-cancer study had been confirmed. However, evidence that the void in animal research was finally being filled was at hand.

At the annual meeting of the Bioelectromagnetics Society, held in Salt Lake City between June 23rd and June 27th, 1991, scientists at Health and Welfare Canada, in Ottawa, announced that sixty-hertz magnetic fields and TPA, a chemical known to be a tumor promoter, had been found to act together to accelerate the development of skin tumors in mice. At the same meeting, Dr. Chris Cain, of the Electromagnetic Research Laboratory at the Jerry L. Pettis Memorial Veteran's Hospital, in Loma Linda, California, had reported that ELF fields had been found to accelerate the action of TPA in cancer cells grown in culture. The fact that magnetic fields had been shown to promote the development of cancer in both live animals and cell cultures was considered to be extremely important by many researchers in the bioelectromagnetic field. In a headline story that appeared in the July/August issue of *Microwave News,* Slesin quoted a number of them, including two members of the Scientific Advisory Board subcommittee (the same subcommittee whose forthcoming review of the E.P.A. report would assert that there was insufficient evidence from animal and cell experiments to infer that power-line magnetic fields could be carcinogenic). One of the subcommittee members was Craig Byus, an assistant professor of chemistry at the University of California at Riverside, who told Slesin that the Canadian animal experiment was "potentially highly significant." The other subcommittee member was Bary Wilson, a molecular biologist at the Battelle Pacific Northwest Laboratories, in Richland, Washington, who declared, "If this study continues to be positive, it will change the playing field."

Further indication that animal research might play a crucial role in defining the power-line health hazard came in September, when it was learned that scientists at the Oncology Research Center of the Republic of Georgia's Ministry of Health and Social Security had conducted experiments showing that power-frequency electromagnetic fields had dramatically increased the incidence and enhanced the speed of development of mammary gland tumors that had been induced in rats by a chemical carcinogen called nitrosomethyl urea. The Georgian researchers emphasized the importance of their study by pointing out that breast cancer was the most common form of cancer and one of the leading causes of death in women, and they warned that exposure to power-frequency fields found in households might lead to an increased incidence of this malignancy.

CHAPTER FOURTEEN

Some Unsatisfactory Answers

ON SEPTEMBER 18TH, the members of the Slater Electromagnetic Field Study Subcommittee held their second meeting at the school. Like the meeting of June 5th, it was presided over by Beauregard, and was attended by representatives of the county health department, the school district, the Slater teachers, Parents for a Safe Environment, the Slater PTA, Berryman, DeYoung, and Neutra. DeYoung was accompanied by three P. G. & E. engineers, and Neutra, for his part, had brought along a twelve-page draft report on the progress of his investigation of the cancer cluster. The P. G. & E. engineers presented a report on some twenty-four-hour magnetic-field measurements that the utility had taken at Slater from August 28th to August 29th, and from September 4th to September 5th. The measurements had been made with Emdex gaussmeters that had been placed two feet above the floor in the center of Berryman's former classroom in Pod A, which was the nearest room in the main school building to the transmission lines; at the same height in Room 20 in Pod D, which was the farthest room from the

lines; and at ground level in the sandbox of the kindergarten playground on Emerson Avenue.

According to the P. G. & E. engineers, the average magnetic fields that were recorded in Berryman's former classroom during both periods were not appreciably stronger
than those recorded in Room 20, which ranged from .09
milligauss to 1.14 milligauss. During both twenty-four-hour
periods, fields of about four milligauss were measured at
the kindergarten sandbox next to Emerson Avenue. When
Berryman pointed out that the measurements taken in
Room 20 made little sense, and that the magnetic fields recorded in her classroom were consistently lower than the
fields she had measured there in June, the engineers had
no explanation, but they acknowledged that the Emdex
meter at the sandbox had given erroneous readings, and
they indicated that other erroneous readings were possible.

Next, Neutra described the study that he and his
colleagues in the Department of Health Services were
planning to conduct of the magnetic-field exposure of
teachers and children in California schools with varying
proximity to power lines. He said that he was in the process
of designing a protocol for this study, which he expected to
begin in the spring of 1992, if money were made available,
and he requested that the staff at Slater consider participating in it. He went on to explain that it would compare
the magnetic-field exposures of teachers and children in
schools near power lines with the exposures of teachers
and children in control schools that were not situated near
power lines.

"When Dr. Neutra mentioned the need for control
schools, I thought it would be appropriate for me to tell
him what I knew about the McCardle School," Berryman
recalls. "McCardle is an elementary school about four or
five miles east of Slater, and it is, pod for pod, Slater's iden-

tical twin. It was built during the late nineteen-seventies from the same set of plans as Slater, and it has precisely the same layout. It is even supposed to have the same wiring and plumbing. In fact, just about the only difference between McCardle and Slater is that there are no high-voltage or high-current power lines near McCardle. In early August, I had gone over there with Karin Nora, a fellow teacher at Slater, and we had spent a couple of hours taking magnetic-field measurements with an Emdex borrowed from P. G. & E. Two of the rooms in Pod A were locked, but the ambient fields were measured in the three other rooms of Pod A and in the five rooms of Pod B were rarely over half a milligauss, and were usually less than that. The levels we measured in the ten rooms of Pods C and D were almost identical to those we measured in Pods A and B. The only places we got high readings at McCardle were when we held the gaussmeter next to three-way light switches, against walls that concealed electrical wiring, and directly beneath low-hanging fluorescent lights. When I told Dr. Neutra about the McCardle School, I gave him a report that Karin and I had written, which described the measurements we had taken there room by room. I then suggested McCardle might make an ideal control school. After all, McCardle was identical to Slater in virtually every respect, except that the ambient magnetic fields in Pods A and B at McCardle were three to six times less strong than those in Pods A and B at Slater. Wouldn't it make sense, therefore, to compare the cancer rate among the teachers who had worked at McCardle with the rate among the teachers who had worked at Slater? For some reason, Dr. Neutra appeared not to think so. In any event, he responded to my suggestion in a very noncommittal manner, saying that it might be something he would consider in the future."

When the subcommittee meeting adjourned, Neutra and

Pamela Long, an industrial hygienist from San Francisco, who had been hired by a free-lance television producer to take magnetic-field measurements at Slater earlier in the month, went through the school to locate some "hot spots" that Long had found in the walls and floor of various rooms in Pods C and D, and in the all-purpose room. Because a question had been raised as to whether the hot spots might be caused by stray currents from sources outside the school, Neutra arranged to have the main breaker switches at Slater shut off. He and Long then remeasured the locations at which hot spots had been detected, and found that they had disappeared — proof that they were not created by stray currents but by electrical wiring.

When Neutra and Long took measurements in the classrooms of Pods A and B, they were unable to find any significant hot spots. However, with the power to the school shut off, they recorded ambient fields of between two and three milligauss in Berryman's old classroom — proof that relatively strong magnetic-field emissions from the high-voltage transmission lines on Emerson Avenue were extending into the south side of the building. (This was the only time that Neutra or any other member of the Department of Health Services measured magnetic fields that were being given off by the lines.) Further evidence of these emissions came when Long measured a magnetic field of nine milligauss at the kindergarten sandbox near Emerson Avenue — more than twice the level that had been reported by P. G. & E. at the subcommittee meeting.

When Berryman got home that evening, she read the progress report that Neutra had written on his investigation of the Slater cancer cluster. In the report, Neutra noted that he and his staff had confirmed eight cases of invasive cancer out of a total of twenty possible malignancies that had by then been reported to have occurred among the one hundred and thirty-nine teachers and teacher's aides who were known to have worked at the school at one time or another since 1972, and among the six office and kitchen staff members who Neutra estimated had been employed there during the same period. (A second case of melanoma was reported to have developed in a school maintenance worker, but it had not yet been verified by Anna Phillips.) According to Neutra, most of the other reported cases had proved to be conditions that were nonmalignant, among them the brain-tumor meningioma, which has been linked in some reports with exposure to power-line magnetic fields; four cases of cervical dysplasia, which is a usually precancerous abnormality in the cells of the lining of the cervix; and three cases of benign tumors or cysts of the breast. Neutra went on to say that six of the confirmed invasive cancers — two each of the breast, the uterus, and the ovary — were biologically related because they had developed in the reproductive system. In addition, he acknowledged that there was "some epidemiological evidence relating several of the types of invasive cancer to electromagnetic field exposure." In this connection, he mentioned a study in which women living near high-current power lines had been found to be experiencing a significant increase of cancer of the breast and the uterus. He also noted that three studies had found elevated risks of breast cancer in men occupationally exposed to electromagnetic fields from power lines and other sources.

In evaluating the incidence of cancer at the Slater School, Neutra employed standard epidemiological and statistical methodology. First, he added up the total years of employment of the one hundred and forty-five teachers, teacher's aides, and staff members, using the information that Berryman and her colleagues had culled from the year-books. This gave him the number of person-years of observation accumulated by the Slater teachers and staff since they had started working at the school. He then analyzed the number of person-years in terms of National Cancer Institute statistics showing the annual rate of invasive cancer in American women between the ages of forty and forty-four — the age group encompassing the average age of the teachers and staff at Slater — to be two hundred and sixty-eight per one hundred thousand women. This enabled him to calculate that 4.2 cases of cancer could have been expected to occur among the Slater teachers and staff members over the nineteen-year period. (Neutra chose not to use data collected by the California Tumor Registry, which set the annual rate of invasive cancer among women between the ages of forty and forty-five who lived in the Central Valley region of the state, where Fresno is situated, at two hundred and five cases per hundred thousand women; according to the Central Valley statistics, only 3.2 cases of cancer would have been expected to occur among Slater teachers and staff members over the nineteen-year period.)

Using a standard statistical method known as the confidence interval, Neutra estimated that twice the number of observed cancers could have occurred by chance. Employing at that point a statistical formula called the Poisson distribution, which is used to assess the likelihood of rare events, he estimated that eight or more cases of cancer could be expected to occur among the teachers and staff members of forty-five out of every thousand schools like Slater. He went on to estimate that as many as a thousand out of a total of eight thousand schools in California might be situated near power lines — a calculation for which he offered no substantiation — and he concluded that "clusters like this in schools near power lines could well occur by chance." He also noted that the teachers and teacher's aides who had developed cancer had "spent most of their time in classrooms located closest to the nearby transmission lines," and said that this presented "an issue we will evaluate further."

Berryman's reaction to Neutra's findings was one of frustration. "First of all, I couldn't understand why he chose to base his conclusions on the expected rate of cancer for the whole teacher population at Slater, when all of the cancer had occurred among teachers and staff members who had worked in Pods A and B and in the nearby administrative area," she said recently. "Second, I thought that his estimate that one out of eight schools in California might be near power lines was inappropriate, unless he really possessed evidence indicating that one thousand out of the eight thousand schools in the state are as close to power lines as Slater is to the high-voltage transmission lines on Emerson Avenue. One thing I know for sure: Dr. Neutra will not find six out of the fifty-two schools in Fresno to be as close to major power lines as Slater is. However, the kindergarten of the Wilson Elementary School, at the corner of Ashland and Hughes, sits within forty feet or so of a high-current feeder line that gives off strong magnetic fields, and two of the four teachers who taught in it over the past twenty years developed cancer. One of them, a former sorority sister of mine, who worked there for eighteen years, died last year of brain cancer. The other developed breast cancer after working there for fifteen years. As for the conclusion of Dr. Neutra's report, I couldn't understand why he had failed to address the central question we had raised six months earlier at the March 14th school board meeting — namely, why had all the cancers occurred in teachers and staff members who worked on the side of the Slater school that was closest to the high-voltage transmission lines on Emerson Avenue? After all, that was why the Slater teachers and parents had insisted that Pods A and B be evacuated to begin with, and why the school board had agreed back in May to evacuate them."

Berryman went on to say that a week or so after Neutra had submitted his report to the subcommittee she met to review it with several colleagues from the committee who were concerned that the school board might try to use the report to reopen Pods A and B. "During October, we met several times to discuss the report at greater length and formulate questions for Dr. Neutra about some of the conclusions he had reached," she said. "Meanwhile, he sent us a letter about the measurements that he and Pamela Long had taken at Slater on September 18th. Most of the letter had to do with where the hot spots were located, and how shutting off the power to the school had shown that they were caused by electrical wiring in the building. In the final paragraph, however, he mentioned in passing — his exact words were 'an incidental note' — that he and Long had measured ambient magnetic fields of nine milligauss at the kindergarten sandbox, and of between two and three milligauss in my old classroom. He went on to point out that these readings were higher than those taken by P. G. & E. at the same sites in late August and early September, and that if they correlated with the utility's power-line load data for September 18th the correlation would underscore the importance of predicting the strength of power-line magnetic fields with a computer model, and thus dispense with the necessity of having someone actually measure them. At that point, it became clear to me that Dr. Neutra and I were approaching the situation at Slater from entirely different perspectives. He was interested in the technical aspects of how to measure the magnetic fields, while I was wondering what harm they might have done to me and my fellow teachers and our pupils in Pods A and B, who had been exposed to them day after day, week after week, year in and year out."

Neutra was unable to attend the next subcommittee meeting, which was held at Slater on October 30th, but he participated in it over a speakerphone system that was set up in Pod C by officials of the school district. "The speakerphone hook-up was a disaster," Berryman said later. "Dr. Neutra's voice kept fading in and out, and we could only hear fragments of the answers he gave to our questions about his draft report on the cancer cluster. During one of the more intelligible segments, he recommended that a man named Karl Riley, who had a magnetic-field measuring firm in Berkeley, come and take readings at Slater. He said that Riley had found and corrected some problems in the fluorescent lighting in his — Dr. Neutra's — office, and that the correction made him feel better about working there, and he suggested that Riley might be able to do the same with regard to some of the hot spots that had been measured at our school." Neutra did not mention something he had acknowledged publicly several months earlier — that the fluorescent lights in his office had been exposing him to a field of two and a half milligauss, which was about the strength of the field that he and Long had recorded in Berryman's old classroom.

"We agreed that Riley should conduct a magnetic-field study at Slater, but we made it clear that in addition to investigating hot spots created by wiring in the floors of some of the classrooms we wanted him to measure the background magnetic fields in Pods A and B which were being given off by the transmission lines on Emerson Avenue. After all, it was the power-line magnetic fields, not hot spots, that had been associated with cancer in the epidemiological studies. We had already informed P. G. & E. in writing that we were concerned about the large discrepancy between the readings of four milligauss or so that the utility had taken at the kindergarten sandbox in late August and early September and the level of nine milligauss that had been measured there by Pamela Long on September 18th. For this reason, we started the October 30th meeting by asking Thom DeYoung to explain the reason for this discrepancy, and to tell us why the twenty-four-hour readings that P. G. & E. had taken in my old classroom in Pod A bore little relation to the amperage load the utility had recorded during the same period on the high-voltage transmission lines. DeYoung said he could not explain the discrepancy or the lack of correlation. In fact, he and a P. G. & E. engineer who accompanied him to the meeting claimed that the loads on the transmission lines were constantly fluctuating, and that it was not possible to use them to predict the strength of the magnetic fields that were being emitted."

Disappointed with the results of the October 30th meeting, Berryman, Charleene Conley, and Lynn Stenson, the president of the Slater Parent-Teacher Association, met on November 13th and November 18th and drew up a list of ten questions they wished Neutra to answer regarding his draft report. Chief among them was why Neutra had used all one hundred and forty-five teachers, teacher's aides and staff members in assessing the expected rate of cancer at Slater, when all of the reported cancer cases had occurred among people working in the ten classrooms of Pods A and B. (Actually, two of the confirmed cancer cases had worked in a part of the administration area that was immediately adjacent to Pod A.) Berryman and her colleagues also wanted to know whether the cancer cases shared any biological similarities, whether they shared any known causes, and how many schools in California were as close to high-voltage transmission lines as Slater. In addition, they informed Neutra that because exposure to power-line magnetic fields had been linked to childhood leukemia, they felt strongly that the Department of Health Services should conduct a thorough study of the incidence of cancer in children who had attended Slater.

Berryman sent Neutra the list of ten questions and the request for the childhood-cancer study on November 22nd. On the same day, Riley came to Slater with Pamela Long to conduct the magnetic-field survey that Neutra had recommended. Four days earlier, Neutra had sent a letter to the school district, explaining that because of budget cuts and Governor Pete Wilson's veto of a bill requiring the Public Utilities Commission and the Department of Health Services to continue their research and education programs regarding the potential health hazards of exposure to electromagnetic fields, the Department had been forced to cut back drastically on its activities. "We have cancelled presentations scheduled for next year, turned down all requests for speaking engagements, and cancelled all plans for a 1992 workshop," Neutra wrote. On November 24th, he informed Fresno School District officials that he would be unable to attend a subcommittee meeting that had been scheduled for November 27th, but would participate once again from his office in Berkeley via the speakerphone system.

"The school district people promised that this time the speakerphone system would work much better, but in fact it was even worse than before," Berryman recalls. "Dr.

Neutra's voice kept fading in and out, and there were whole periods of time when we couldn't hear him at all. The reception got so bad that he tried to use Riley, who was participating in the conference call from another site, to relay his words to us. That didn't work because we couldn't hear Riley, either. In the end, we had only fragments of Dr. Neutra's conversation to try to piece together. I believe he said that six of the eight invasive cancers to be, as he put it, 'second cousins,' because they affected the reproductive system, but that he didn't think they shared any common causes. As for how many California schools were as close to power lines as Slater, he admitted that he didn't know. When we pressed him about our request that the Department of Health Services study the incidence of cancer among former students at Slater, he told us that there were no funds available for such a study."

Berryman went on to say that, because of the poor speakerphone reception, she, Conley, and Stenson were as dissatisfied with the results of the November 27th meeting as they had been with the results of the October 30th meeting. "Two and a half months had passed since Dr. Neutra had given us his draft report, and we still had not been given satisfactory answers to many of the questions we had raised about it," she said. "The situation got even murkier, two weeks later, when officials of the school district sent us Karl Riley's report of his survey. In his report, Riley said that the purpose of his visit to Slater was to investigate and reduce high magnetic fields generated by building wiring — hot spots that had been identified in earlier surveys — and he appeared to think that the classrooms at Slater had been vacated because of these hot spots. That, of course, was not the case. The classrooms in Pods A and B had not been vacated because of hot spots in wiring, most of which had been found in Pods C and D, but because of the high rate of cancer among the teachers who had worked in Pods A and B, and because of the proximity of these two pods to the high-voltage transmission lines on Emerson Avenue. Elsewhere in his report, Riley claimed that fixing wiring in Rooms 6 and 7 of Pod B would reduce the magnetic-field levels in those rooms to the background levels given off by the transmission lines, which he said were under one milligauss on the day of his visit. He went on to suggest that if high fields from faulty wiring in Rooms 6 and 7 were the reason for moving children out of Pod B, there was no longer any reason not to move them back. Our hearts sank when we read that, because none of us who accompanied Riley on the day of his visit, including a P. G. & E. engineer, had observed him taking any background readings, nor had he listed any background readings in his report, and because school district officials, who had hired Riley on Dr. Neutra's recommendation, had made no secret of the fact that they wanted to re-open Pods A and B as soon as possible. It was a year almost to the day since Amy Alexander had appeared in the teacher's lounge and alerted us to the hazard posed by our working so close to the transmission lines, and we began to wonder if we were being given the runaround."

PART THREE

COVER-UP AND CONFRONTATION

CHAPTER FIFTEEN

Working Together to Protect the Public Health

U NKNOWN TO BERRYMAN and her colleagues, one reason Neutra had had so little time for the Slater School subcommittee during the autumn was that he had been named to the California Electromagnetic Fields Consensus Group — a seventeen-member committee that had been set up by the Public Utilities Commission as a part of its investigation into the potential health hazard of power-line magnetic fields — and had been meeting with the group in day-long sessions at least once a week since early October. The Consensus Group had been given one hundred and twenty days by the Public Utilities Commission to propose priorities for utility-financed research, and to recommend interim action that the Commission and the electric-utility companies might take to mitigate the health risks of electromagnetic-field exposure, until scientific evidence concerning the power-frequency hazard provided better direction for public policy. In addition to Neutra, the group included five utility-company officials — among them representatives of Pacific Gas & Electric, Southern California Edison, and San Diego Gas & Electric; three representatives from public-interest and environmental groups; two ratepayer advocates; two members of the International Brotherhood of Electrical Workers; an environmental consultant; a member of the California Directors of Environmental Health; a member of the California Energy Commission; and the dean of the San Diego State University's College of Science.

Many observers believe that the utilities joined the Consensus Group because they were alarmed about the provisions of a bill that had been introduced in the California State Legislature by Senator Herschel Rosenthal, chairman of the Senate Committee on Energy and Public Utilities. Rosenthal's bill had passed both houses of the legislature during the summer of 1991, but had been vetoed by Governor Pete Wilson, on October 14th, as the Consensus Group was beginning its deliberations. The bill had called for the Commission and the Department of Health Services to study implementing a policy of prudent avoidance in order to reduce exposure to electromagnetic fields; it would also have required the utilities to develop programs in response to customer requests for electromagnetic-field measurements, and to pay a one-time fee of up to seven million dollars to finance these programs.

Whatever the motives of the utilities for joining the Consensus Group, it had soon become evident that they saw it as a vehicle for what one member of the group has called "risk containment." In May, while objecting to a proposal that would make utilities responsible for reducing magnetic-field emissions from their power lines and substations to one milligauss or less on private property, the San Diego Gas & Electric Company had told the Public Utilities Commission that "absolute proof ought not to be required before public policy dictates mitigation of any perceived public health problem," and that in the case of electromagnetic fields "science does not yet know whether there is a problem, or exactly what the problem might be." Since several studies had found elevated risks of cancer among children exposed at or above two to three milligauss in homes near high-current distribution lines, and since San Diego Gas & Electric was about to be named as a defendant in a landmark personal-injury lawsuit brought in behalf of a four-year-old child who had developed cancer after being exposed to magnetic fields of between four and a half and twenty milligauss that had been given off by one of the utility's high-current distribution wires, the company's suggestion that no one knew what the problem might be seemed self-serving, to say the least. Additionally self-serving was San Diego Gas & Electric's insistence upon equating the biological effects of power-line emissions with those accompanying acts of God. The proposal to reduce magnetic field to one milligauss "ignores that life is a process of balancing risk with the cost of risk avoidance," its attorney, E. Gregory Barnes, wrote. "For example, many people choose not to live in California because of the risk of an earthquake. Such decisions are personal, and no one suggests that the state ought to insulate people from all potential risk associated with living in a seismically active area. Sound public policy addresses seismic risk with building codes and other public safety measures, including prediction research. But, as with earthquakes, the state should not insulate people from *all* possible risk [that] some associate with EMF from utility facilities."

Southern California Edison and Pacific Gas & Electric also took the position that the evidence associating power-line emissions with the development of cancer did not warrant significant action by the Public Utilities Commission. In

April, P. G. & E. attorneys had told the Commission that "scientific research has not demonstrated that 60 hertz electric and magnetic fields cause adverse health effects." To support this contention, they quoted from a report that had been submitted in 1989 to the California State Legislature by officials of the Public Utilities Commission and the Department of Health Services, who claimed that "we do not know which components, if any, of electric power utility operations pose significant health hazards." With this as a springboard, the utility's attorneys recommended that the Commission not adopt a standard for exposure to electromagnetic fields "because there is currently no scientific basis for determining the nature of exposure which should be mitigated, if any, or what kind of mitigation is appropriate and feasible." Apparently confident that the Department of Health Services would not soon find otherwise, they went on to recommend that it be given the task of managing electromagnetic-field research.

Early in their deliberations, the seventeen members of the Consensus Group agreed to a proposal put forth by staff members of the Public Utilities Commission that all consensus recommendations had to be unanimously approved, which meant that any single member could veto any recommendation. They also established a protocol allowing for nonconsensus views to be included in their final report. Since the threat posed by Rosenthal's bill had been removed by Governor Wilson's veto, and since the three major utilities were by then either involved in defending personal-injury and property-devaluation lawsuits involving power-line emissions, or in preparing to defend against such litigation, it was a foregone conclusion that utility representatives on the consensus group would not agree to any proposals that might be construed as admitting that power-line magnetic fields posed a hazard to human health. In short, the unanimous approval requirement meant that only the blandest of recommendations could possibly achieve consensus.

In their final report, a seventy-six-page document that was issued on March 23rd, 1992, the Consensus Group members showed themselves to be divided on almost every major issue regarding the power-line health hazard. After agreeing that "some epidemiological studies report a statistically significant association between high-current wiring configuration and childhood cancers," and that "other epidemiologic studies report statistically significant associations between certain 'electrical occupations' and leukemias, lymphomas, brain cancer, and male breast cancer," they found themselves unable to agree on how to interpret the results of these studies. They concurred that "risks suggested by certain research findings are potentially serious and have important public health implications," but could not agree on how to deal with the potential risks — for example, whether it was reasonable to impose an involuntary potential risk on those who live in close proximity to power lines.

At no time during their five-month-long deliberations did the members of the Consensus Group engage in any detailed discussion of the three-hundred-and-sixty-seven-page report that had been issued a year earlier by staff scientists of the Environmental Protection Agency, who had initially concluded that power-line magnetic fields were a "probable" cause of cancer. Indeed, the only significant mention of the E.P.A. report appears to have occurred when Neutra and Jack Sahl, a research scientist working for Southern California Edison, told the rest of the group that it had been criticized by the Scientific Advisory Board, and that

the epidemiological evidence it described as linking exposure to magnetic fields with the development of cancer
was inconclusive. Sahl, who had previously assured the residents of Santa Barbara that Montecito Union was "just a
normal school in terms of electromagnetic fields," had published an article in the October 1991 issue of the Health
Physics Society's monthly newsletter, in which he suggested, as Neutra had in the Department of Health Services'
1990 pamphlet, that epidemiological studies demonstrating
an association between EMF exposure and childhood cancer were balanced by other "equally well-done" studies
that did not show the association. The fact that no such balance existed could have been easily ascertained by anyone
who took the time and trouble to read the E.P.A. report, or,
for that matter, pages fifty-two and fifty-three of the Consensus Group's report, where the following sentence could
be found: "A recent table summarizing fifty-one (51) epidemiological studies conducted on EMF surrogates and
cancer indicated that 28 studies (or 55%) reported a statistically significant risk, fifteen studies (29%) reported elevated but nonstatistically significant risk, and eight studies
(16%) reported no association." (Ironically, a footnote to
this sentence reveals that the table summarizing the fifty-
one epidemiological studies appeared in a 1991 Department of Health Services' handbook that was distributed to
people attending electromagnetic-field workshops in Oak-
land and Newport Beach, which were conducted by none
other than Neutra.) As for the willingness of the Consensus
Group to accept the assertions of Neutra and Sahl about the
epidemiological evidence, one of its members explained
that "the reason none of us challenged what Raymond and
Jack said about the studies was that they were the only epidemiologists in our group, and we looked to them as the
experts."

Once they allowed themselves to be persuaded that the evidence indicating power-frequency magnetic fields as a carcinogen was either flawed or inconclusive, the members of the Consensus Group encountered little difficulty in reaching unanimous agreement on proposals set forth by Neutra, calling for the Public Utilities Commission to establish an independent California research program that would be managed by the Department of Health Services. Curiously, the Department's 1990 pamphlet, in which Neutra and his colleagues had done little more than reiterate the essential elements of the utility industry's position regarding the power-line health hazard, was never discussed when these recommendations were deliberated by the Consensus Group members, who also agreed that the Department should develop a repository for information about power-line cancer clusters. Developing a repository for information about cancer clusters was one thing, but doing something about them was quite another, as the members of the Consensus Group soon learned when they found themselves unable to agree that the Public Utilities Commission should work with the Department of Health Services to develop a program for responding to reported power-line cancer clusters.

Since the final report of the Consensus Group did not identify the positions taken by individual members of the group, or how the members had voted on the various issues before them, it provided no record of accountability. For the most part, however, the proposals that failed to achieve consensus were supported by the representatives of the citizen-action and environmental groups, and were opposed by representatives of the three major utilities and/or by the representative of a group called Toward Utility Rate Normalization, who voted against almost every measure that might increase the cost of electricity for ratepayers. Among the nonconsensus proposals was one that called for evaluating the existing work practices of utility, telephone-company, and cable-television employees, who work in close proximity to power lines, with the goal of reducing their exposure to magnetic fields. Strangely enough, this proposal was not supported by the two representatives of the International Brotherhood of Electrical Workers who sat on the Consensus Group. They favored a utility-sponsored recommendation that only new work practices be evaluated, and then only with the goal of implementing no-cost or low-cost measures to reduce worker exposure.

The failure of the IBEW to support the proposal about existing work practices seems puzzling in light of the fact that during the previous ten years more than two dozen epidemiological studies had been conducted and published in the peer-reviewed medical literature in the United States and around the world, showing that power linemen and other workers whose occupations expose them to electromagnetic fields were developing and dying of leukemia, lymphoma, and brain cancer at rates significantly higher than those observed in unexposed workers. (Indeed, a recent investigation conducted by researchers at the University of Southern California School of Medicine, in Los Angeles, had shown that electricians, electrical engineers, and other workers exposed to electromagnetic fields in Los Angeles County were developing brain cancer at more than ten times the expected rate.) However, on June 28, 1991, J. J. Barry, the IBEW's international president, had released a cautiously worded statement declaring that there had been "no proven excess occupational disease or any fatalities to IBEW members which to our knowledge can be directly attributable to either ionizing or nonionizing radiation." Barry went on to say that the union would adopt a wait-and-see attitude regarding the power-line hazard, until the results of further studies became known. Just how keen top IBEW officials were about looking into the hazard can perhaps best be judged by the fact that on two occasions during the previous ten years they had declined the offer of Dr. Samuel Milham, Jr., a physician and epidemiologist with the Washington State Department of Social and Health Services, who had conducted several pioneering studies of disease in electrical workers, to investigate at no cost to the union the mortality experience of IBEW members, in order to determine whether their exposure to power-line magnetic fields was placing them at an increased risk of developing cancer.

Thanks to opposition by the utilities, the members of the Consensus Group could not agree on a proposal calling for the utilities to issue annual reports on the status and cost of mitigation strategies that were presently available and technologically feasible for lowering power-frequency emissions. Among these strategies were reverse phasing of conductors; reconfiguration of lines to reduce magnetic-field emissions; placing power lines on higher poles and towers; employing wider rights-of-way; balancing currents on distribution lines; burying power lines so that currents were balanced and magnetic fields cancelled out; and avoiding the placing of transformers adjacent to office space in large buildings. Opponents of the proposal described it as "unduly burdensome," and insisted that since scientists had been unable to pinpoint exactly which components of electromagnetic-field exposure might pose a health risk, any decision to mitigate exposure should be carried out on a voluntary basis.

As for what to do about the power-line hazard in

schools, the Consensus Group members found themselves unable to agree to a proposal calling for utilities to "avoid siting new transmission lines, distribution lines, or substations adjacent to sensitive receptors such as schools and child care facilities." Nor could they agree on a recommendation calling for the Public Utilities Commission to adopt specific procedures for utilities to follow in responding to the concerns of parents, teachers, and school administrators at schools in which cancer clusters believed to be caused by electromagnetic fields had been reported. Group members who supported these proposals pointed out that the London-Peters study had shown a two-and-a-half-fold increase in leukemia in Los Angeles children living in homes close to high-current lines, and that the California Department of Education now required that new schools be situated farther away from high-voltage transmission lines than in the past. They also drew attention to the fact that the Swedish National Energy Administration had recently recommended against placing schools, kindergartens, and playgrounds near transmission lines. Utility representatives who opposed the proposal declared that "to suggest that particular people or types of locations such as schools and child-care centers be singled out for special siting recommendations sends a false message to the public and could unnecessarily alarm people." They went on to claim that "scientific evidence has not clearly indicated that children are at any greater risk than adults," and that keeping power lines away from schools and child-care centers could prove to be "potentially costly."

In the opening statement of their report, the members of the Consensus Group put the best possible face on the results of their deliberations by declaring, "We believe that the consensus process that produced this report will enable

the California Public Utilities Commission to better understand the EMF issue and develop a policy based on the
views of all interested parties." They went on to say that
"we know that California's utilities, customers, and others
can work together to protect public health by setting sound
policy based on the best scientific information available."
Not surprisingly, the Public Utilities Commission was only
too happy to quote these optimistic assessments in a two-
and-a-half-page news release that accompanied the publication and distribution of the final report. In fact, however,
the rosy expectations of the Consensus Group bore little re-
semblance to the actual behavior of the Commission and
the utilities regarding the power-line health hazard.

An example of how officials of the Commission and the
utilities were really conducting themselves occurred that
same winter of 1992, in Bakersfield — a city of almost two
hundred thousand inhabitants that is situated at the southern end of the San Joaquin Valley, about one hundred and
twenty miles south of Fresno. Early in January, some residents of Campus Park, a subdivision in the southwestern
section of the city, woke up one morning to find that Pacific
Gas & Electric had begun to erect a series of ninety-foot-tall
steel transmission towers along the edge of a transporta-
tion-corridor right-of-way that ran within thirty to fifty feet
of approximately seventy homes. They soon found out that
the towers were being built to carry three pairs of hundred-
and-fifteen-thousand-volt transmission lines that were supposed to supply electricity for some twenty-four thousand
residences and businesses that were expected to be constructed in Bakersfield in the future. On January 15th, fifteen residents of Campus Park met with Paul Elias, the
manager of P. G. & E.'s Kern County Division, at the home
of Kathleen Marzolf, who had organized a group called

Concerned Citizens of Campus Park. When Marzolf and her
neighbors told Elias that they were worried about the cancer hazard posed by the magnetic fields that would emanate
from the new transmission lines, he tried to placate them
by offering to plant trees that would hide the lines from
view.

On January 25th, a P. G. & E. engineer informed one
hundred and fifty Campus Park residents who attended a
meeting at the subdivision's Sing Lum School, and were
upset that they had not been notified about the proposed
power lines, that the utility was not required by the Public
Utilities Commission to give advance notification about the
construction and environmental impact of any transmission
line that carried less than two hundred thousand volts.
When people in the audience expressed fear that the presence of high-voltage lines would devalue their property, a
P. G. & E. spokesman told them that there was no consensus
in California about whether that would be the case. He
either did not know about or chose not to acknowledge the
landmark 1988 case of *San Diego Gas & Electric Company
v. Donald L. Daley*, in which a California Court of Appeals
affirmed a trial-court verdict awarding more than a million
dollars in damages to a landowner, who alleged that the
value of his property had been diminished by a high-
voltage transmission line that the utility had built across it.
The appellate court had also upheld the trial court's decision to admit evidence about the potential health hazard
posed by power-line electromagnetic fields and its adverse
effect on property values.

The chief speaker at the January 25th meeting was Cath-
erine Moore, the director of P. G. & E.'s electromagnetic
field program, who, as it turned out, had served for several
weeks during the autumn as the utility's first representative

on the Consensus Group. According to Marzolf and other Campus Park residents who attended the meeting, Moore began her talk by describing some serious health problems of her own. "When she thought she had engaged our sympathy, she pulled a hair dryer out of a suitcase she had brought with her and tried to tell us that the magnetic fields it gave off would be greater than those given off by the new transmission lines," Marzolf recalls. "At that point, people in the audience reminded her that nobody holds a hair dryer to his head for twenty-four hours at a time, and that she should know better than to make such a statement. That seemed to rattle her a bit, because she put the hair dryer away and admitted that no one could deny that power-line magnetic fields might pose health problems. She went on to say, however, that studies of the power-line hazard were inconclusive, and that nobody knew what level of magnetic-field exposure was safe or unsafe. She also claimed that if the transmission line were buried, we would be exposed to stronger magnetic fields than would be given off if the wires were strung on the ninety-foot towers, because our homes would be closer to the buried wires, and because the ground does not provide a shield from the magnetic fields. We didn't know enough to correct her on that one, because at that point we hadn't done enough reading to know that if transmission lines are buried close together in steel casings filled with oil, the magnetic fields they emit can be sharply curtailed."

On March 18th, five days before the Public Utilities Commission released the report of the Consensus Group, another meeting at the Sing Lum School was addressed by George Hersh, an environmental program manager for the Commission, who, as it happened, had served as a liaison person to the Consensus Group, and had attended most of its meetings. The Commission sent Hersh to Bakersfield at the request of Marzolf and other members of Concerned Citizens of Campus Park, who had filed a complaint with the Commission, objecting to P. G. & E.'s plan to construct high-voltage lines in their community. Marzolf remembers that a week or so before Hersh arrived, she telephoned him at the Commission's headquarters, in San Francisco, and told him that the members of her organization wanted him to discuss the power-line hazard in detail, and that they especially wanted to know what constituted a safe distance to live from the new transmission lines, and what would be a safe level of magnetic-field exposure for themselves and their children. "Hersh led me to believe that he would address these questions fully, but when he got up to speak on the night of March 18th he sounded like a P. G. & E. employee," Marzolf said recently. "He told us that he would like to give us solid answers to our questions, but that scientific evidence about the power-line health hazard was inconclusive, so there were no solid answers. He said that the studies of childhood cancer that had been conducted were unreliable, and he went on to suggest that while some people might be insensitive to electromagnetic fields, there was no real proof of this. At one point, he took a gauss-meter and walked around the room with it, and told us that he was getting different readings in different places. 'We know there's a danger zone somewhere, but we just don't know where,' he said. Then he tried to tell us that the magnetic fields given off by the new transmission lines would cancel each other out. At that point, I grabbed the microphone away from him and asked him to explain how that could possibly be so. He did some hemming and hawing, before admitting that the fields wouldn't cancel each other out completely. By then, a lot of people had got up and left the meeting in disgust. They simply couldn't understand why the Public Utilities Commission, to whom we had gone for help, would send someone down to talk to us like that. When the meeting was over, and Hersh saw me talking to Christine Azzara, a reporter for the *Bakersfield Californian*, he came right over and told her that he had simplified some of his remarks and answers to our questions so that people in the audience might better understand him. Then he admitted that he might have said some things he shouldn't have said, and asked her not to quote him word for word. Later, I listened to an interview he had given to a local radio station and heard him say that it was highly unlikely that power-line magnetic fields could cause cancer. By that time, I realized that we'd been had — first by P. G. & E. and now by the Public Utilities Commission. I felt betrayed. It had never occurred to me that the government would treat ordinary citizens like us with such contempt."

CHAPTER SIXTEEN

Nothing to Break into a Sweat About

MEANWHILE, in Connecticut, officials of the Department of Public Utility Control, the utility companies, and the state's health agencies had allowed 1991 to come and go without taking any effective action to reduce the exposure of unsuspecting citizens to power-line emissions. In December, Don Michak reported in the *Journal Inquirer* that a year after coming into existence the interagency task force had still not undertaken the state-wide study that the legislature had called for, and that state health officials were still unwilling to propose any regulations to control the exposure of the public to electromagnetic fields. Michak quoted Carl Pavetto, of the Department of Environmental Protection, as saying that the reason the task force had decided not to make any recommendations was that "there are a lot of unanswered questions." Just how seriously Pavetto was seeking answers seemed open to question, when he admitted to Michak that he was "not aware" of the London-Peters leukemia study — far and away the year's most widely discussed investigation into the power-line cancer hazard, and the subject of stories in leading newspapers in Connecticut and across the nation.

As in California, however, where public health officials had also failed to take action to protect the public health, the last recourse of the injured citizen would prove to be the law. In the middle of December, attorneys for the Bridgeport law firm of Koskoff, Koskoff & Bieder brought a lawsuit on behalf of Melissa Bullock and her mother against Northeast Utilities and its subsidiary, the Connecticut Light & Power Company. The suit was filed in Connecticut Superior Court, in New Haven, and it alleged that the young woman's malignant brain tumor had been caused by her ten years of exposure to electromagnetic fields given off by the high-current distribution lines that led past her home on Meadow Street from the nearby substation, and that Northeast Utilities and Connecticut Light & Power, the owners of the lines and the substation, had failed to warn her and her mother of the danger posed by these emissions. On January 14, 1992, Koskoff, Koskoff & Bieder filed a similar lawsuit against the same defendants in behalf of Jonathan E. Walston III and his wife, Leah. Walston had recently suffered a recurrence of the meningioma for which he had been operated on in 1979. His lawsuit also alleged that Northeast Utilities and Connecticut Light & Power knew or should have known that the electromagnetic fields given off by the substation and high-current wires on Meadow Street could cause tumors and other serious disease, but had failed to warn him or his family of the hazard.

These landmark actions — the first of their kind to be brought in New England — received considerable attention in legal circles, and were the subject of stories in leading newspapers in Connecticut and Massachusetts. The president of Trial Lawyers for Public Justice — a public-interest law firm in Washington, D.C., which had joined in both cases in behalf of the plaintiffs — said that the lawsuits were of critical importance to other victims of electromagnetic fields, and that the nation's utilities "must stop avoiding the evidence that these power lines are dangerous, and start protecting the public's health." Michael Koskoff, the lead attorney in the cases, told the *Bridgeport Post* that the utilities had known for decades about the health hazards of power-line emissions, but had failed to warn the public. Koskoff, who predicted a protracted court battle, was quoted by the Associated Press as saying that "power companies are going to resist no matter how compelling the facts because they don't want to go to the expense of fixing the problem or compensating the victims." Emmanuel Forde, a spokesman for Northeast Utilities, told the *Boston Globe* that the utility had been following the power-line issue for at least fifteen years, and that "we feel very strongly the scientific evidence is not there to prove that our lines could have caused the health effect." Louis Slesin, of *Microwave News*, pointed out that the power-line health issue had finally ended up in the courts because state and federal health agencies had failed to deal with the problem over the previous twenty years.

While the tragedy of Meadow Street was once again making news in New England, Berryman and her colleagues at Slater Elementary were saddened and distressed by the news that Curtis Hurd had been diagnosed with colon cancer and was in the hospital. The vice-principal, who was forty-three years old, had worked for six years in a tiny office that was only a few feet from Pod A, and was close to the workplaces of four other administrative-staff members who had developed cancer. His illness served to reinforce the skepticism with which Berryman and many of her fellow teachers had come to regard the study that was being conducted by Dr. Neutra and his colleagues at the Department of Health Services.

"We knew that most of the sixty-five teachers and staff members who worked at the school at present had either filled out the illness-report forms or were aware that a cancer study was in progress," Berryman said recently. "What we didn't know was whether Dr. Neutra and his colleagues had got in touch with the eighty-five or so teachers and staff members who had left Slater and were living and working elsewhere. During the next two weeks, Doris Buffo and I telephoned fifteen or twenty of our former colleagues, who were still in the Fresno area, and learned to our astonishment that not a single one of them had been approached. One teacher who had worked for several years in my old classroom in Pod A had developed a nonmalignant tumor of the liver. So we now had good reason to believe that Dr. Neutra and his colleagues had not bothered to investigate the health experience of more than half of the one hundred and fifty people who had worked at Slater during the nineteen years since the school had opened. We were shocked and dismayed by this revelation. In fact, we went from wondering whether we were being given the runaround to realizing that we were being strung along."

Around that time, school-district officials had notified Berryman that another subcommittee meeting had been scheduled for Monday, February 3rd, and that Dr. Neutra, who had previously accepted an invitation to address a luncheon meeting of the Fresno Rotary Club on the same day, planned to attend. In preparation for the subcommittee meeting, Neutra and Dr. Glazer faxed Berryman, Conley, and Stenson a fourteen-page document entitled "Progress Report on Cancer Cases at Slater School," with a four-and-a-half-page appendix containing answers to the questions that Berryman and her colleagues had mailed to Neutra on November 22nd. Both documents arrived in Fresno on

January 31st. The progress report, which was dated January 30th, was almost identical to the one that Neutra had brought to the September 18th subcommittee meeting, except that by this time he and Glazer had confirmed a ninth case of invasive cancer (this was the melanoma in the maintenance worker) and were in the process of verifying the diagnosis of colon cancer in Curtis Hurd. The nine confirmed cancer cases included two breast cancers, two uterine cancers, two ovarian cancers, two melanomas, and a brain cancer; however, this list was incomplete because it did not include Hurd, or a teacher's aide who had developed breast cancer, or the administrative staff member who had developed cervical cancer but had declined to participate in the study for reasons of privacy. As in the previous report, Neutra and Glazer acknowledged that six of the cancers — those of the breast, uterus, and ovary — were biologically related, because they had developed in the reproductive system. They also mentioned the study that had found an increased incidence of breast and uterine cancer in women living in homes near high-current distribution wires, and other studies that had found excess brain and breast cancer, as well as melanoma, in male workers whose occupations exposed them to power-frequency magnetic fields. However, after declaring that most scientists considered these findings to be inconclusive, Neutra and Glazer declared that "there is no one agent that had been solidly documented to cause the various types of cancers observed at the Slater School."

As in the September report, Neutra and Glazer used the National Cancer Institute's estimate that two hundred and sixty-eight cases of cancer could be expected to occur each year in every one hundred thousand American women between the ages of forty and forty-four to determine the number of cancer cases that could have been expected to occur among the one hundred and forty-five teachers, teacher's aides, and staff members who had worked at the Slater School at one time or another since 1972. On this occasion, the two state health officials performed a recalculation that enabled them to revise the expected number of cancer cases among the Slater personnel from 4.2 to 4.5. Once again, they chose not to use the California Tumor Registry's estimate that two hundred and five cases of cancer would be expected each year in every one hundred thousand women between the ages of forty and forty-four who lived in the Central Valley region. Had they used the Central Valley figures, the expected number of cancer cases among Slater teachers and staff members would have been slightly more than three, not the four and a half they had calculated, and this would have brought the cancer incidence among Slater personnel to be nearly three times, instead of twice, the expected rate. And if the total number of people afflicted with cancer had included the teacher's aide with breast cancer, the administrative staff member with cervical cancer, and Curtis Hurd's colon cancer, the incidence of malignancy among Slater personnel would have been approximately three times the national average and four times the average for the region in which the cancers were occurring.

In the summary section of their progress report, Neutra and Glazer concluded that the nine cases of invasive cancer, which they described as being "over twice" the expected average, "could occur by chance." They went on to say that the number of cancer cases they knew about at the present time was "not statistically significant," in terms of the total staff of the school. As for the fact that all of the teachers who had developed cancer had spent most of their time in classrooms in Pods A and B, they said that they had evaluated this issue in the draft appendix that accompanied the progress report.

The draft appendix, which appeared to have been written by Neutra, claimed that none of the Slater teachers or other employees at the school had experienced any unusually strong exposure to power-line magnetic fields. To support this contention, Neutra said that measurements taken at the school showed that the magnetic-field contribution from the power lines to the nearest classrooms was mostly between a half and one and a half milligauss. He was apparently referring to the readings that had been taken by P. G. & E. engineers during the two twenty-four-hour periods in late August and early September, when, by their own admission, at least one of the three Emdex meters they were using had given erroneous readings.

Since Neutra was present at the September 18th subcommittee meeting when the P. G. & E. engineers admitted that their measurements may have been faulty, his use of possibly flawed readings to buttress the contention that no one at Slater experienced strong magnetic-field exposure seems curious — especially in light of the fact that on at least a dozen days between June 10th and July 5th Berryman and a colleague had used Neutra's own gaussmeter to measure background magnetic fields of between one and a half and four milligauss in those same classrooms, and had reported their readings to him. Additionally curious was Neutra's claim on page 5 of the appendix that on one occasion he had measured a magnetic-field level of around two milligauss in a Pod A classroom, when, in fact, he had acknowledged in his letter of October 8th that this reading was between two and three milligauss. The fact that he believed a field of two and a half milligauss to be hazardous was, of course, implicit in his decision to have Riley correct the problems in the fluorescent lighting in his own office, and in his subsequent admission that he felt better about working there, once the magnetic field had been reduced.

In a further attempt to show that transmission lines on Emerson Avenue could not have played a part in the cancer cluster at Slater, Neutra evaluated magnetic-field exposure in various parts of the school campus according to a wiring-configuration code that had been devised more than ten years earlier by Nancy Wertheimer and Ed Leeper, of Boulder, Colorado. Wertheimer, an epidemiologist, had conducted the pioneering study during the late 1970s, showing that children in the Greater Denver area, who lived in homes near distribution wires carrying high current and giving off strong magnetic fields were dying of cancer at about twice the rate of children who lived in homes that were not close to high-current distribution wires. Leeper, a physicist, assisted her in her investigation. The wiring-configuration code they developed was a method of estimating magnetic-field exposure according to how close the residential dwellings inhabited by the children under study were to high-current distribution wires. Wertheimer and Leeper had designed the code to serve as a surrogate for magnetic-field exposure, because at the time they conducted their investigation they were not able to enter homes and take actual measurements. Over the next decade, however, they and other researchers engaged in studying the power-line hazard had come to realize that wiring configuration was a better gauge for estimating magnetic-field exposure over time than spot measurements, for the simple reason that — as DeYoung and the P. G. & E. engineer had acknowledged on October 30th — power-line magnetic fields can fluctuate widely with the time of day and year. The Wertheimer-Leeper wiring code was divided into four categories: a dwelling situated within fifty feet of a high-current distribution or high-voltage transmission line was considered to be a very-high-current home; a dwelling situated between fifty-one and a hundred and twenty-nine feet of such lines was considered to be an ordinary-high-current home; a dwelling situated between a hundred and thirty and a hundred and fifty feet was an ordinary-low-current home; and dwellings situated at distances of more than one hundred and fifty feet were described as very-low-current homes.

Now, faced with the problem of having to explain why so many cancers had occurred among people working on the south side of the Slater School, Neutra, who had previously declared that the wiring code didn't apply to California, decided that it was relevant, after all. To support his use of it, he pointed out that in several residential studies in which the wiring code had served as a surrogate for magnetic-field exposure, excess cancer had been found primarily among children and adults living in very-high-current homes, and, to a lesser extent, in ordinary-high-current dwellings. He then took an aerial site plan of the Slater School and, using its one-inch-to-forty-feet scale, estimated distances of fifty, one hundred and twenty-nine, and one hundred and fifty feet from the middle of the nearest transmission line to various parts of the school grounds, and drew a series of parallel lines across the site plan to mark the boundaries of the four wiring-configuration categories. According to his calculations, none of the campus fell within the very-high-current zone, and only part of the playground and a corner of Pod A fell within the ordinary-high-current zone. The rest of Pod A and all of Pod B and the administrative area — the parts of the school where most of the people who had developed cancer had worked — fell into the ordinary-low-current or very-low-current zones. This enabled him to declare that a majority of the Slater School employees should be classified as occupying the "epidemiologically 'low risk' part of the campus."

If Neutra had considered the magnetic-field levels that had been measured in several of the epidemiological studies in which the wiring-configuration categories were employed as a surrogate for exposure, he might not have been so quick to make this assertion. For example, when Wertheimer and Leeper had conducted a study of adult cancer in Denver a decade ago, they had measured an average magnetic field of only two and a half milligauss in a sampling of very-high-current homes and of only one milligauss in ordinary-high-current homes. As it happened, these data were from the study that Neutra and Glazer had cited in their progress report as showing that women living in homes near high-current wires giving off strong magnetic fields were developing excess breast and uterine cancer; the two health officials had failed to grasp the significance of the fact that the average magnetic fields to which these women had been exposed were of about the same strength as those that had been measured in classrooms on the south side of the Slater School. Moreover, average magnetic fields that were only about half as strong as those measured in the Denver homes were recorded in very-high-current and ordinary-high-current homes in Los Angeles by London, Peters, and their associates at the University of Southern California, who found a significantly increased incidence of leukemia among children living in those homes.

These sets of measurements, together with the fact that

Berryman and her colleagues had recorded background magnetic fields of between one and a half and four milligauss in classrooms in Pod A, ought to have alerted Neutra to the possibility that either he was applying the wiring configuration code incorrectly or that the code itself, which Wertheimer and Leeper had always described as a "very gross method of measuring," might not relate accurately to the magnetic fields given off by the high-voltage transmission lines on Emerson Avenue. Neutra, however, seemed determined to use the wiring code to prove that the transmission lines on Emerson Avenue could not be associated with the cancer that had developed among the teachers and staff members who worked on the south side of the school, and to show that the cancers at Slater must have occurred by chance.

Over the weekend, Berryman read the progress report and the appendix several times. When she examined the aerial site plan, she saw at once that Neutra was mistaken in assuming that the transmission lines followed the direction of Emerson Avenue, and that because of this error the exposure zones he had drawn on the site plan were inaccurate. Since she had never heard of the Wertheimer-Leeper wiring-configuration code before reading the appendix, she telephoned various researchers around the country in an attempt to learn something about it. She found out that it had been designed primarily on the basis of measurements taken near high-current distribution wires, and that its authors had warned against using it without careful evaluation of the configuration in question. She also learned that London and Peters had suggested that the wiring code be refined and improved.

On Monday, February 3rd, Thom DeYoung, of P. G. & E., picked up Neutra at the Fresno airport and drove him to the

Rotary Club in the downtown section of the city, where he gave his talk. Early in his remarks Neutra told a joke about the electric chair, which drew laughter from the two hundred Rotarians who were present, and went on to deliver a thirty-five-minute speech about electromagnetic fields, during which he never mentioned the Slater School or the fact that he had been conducting a study of the incidence of cancer among the employees there. In the course of his speech, he criticized the methodology of a childhood-cancer study he himself had cited to support his use of the wiring-configuration code in the report he had sent to Berryman and her colleagues three days earlier, and declared that the twofold risk of cancer found among children in three studies was "not the kind of thing that I would break into a sweat about." In dismissing the power-line hazard in such glib fashion, Neutra overlooked the fact that the twofold risk found in the childhood studies had undoubtedly understated the problem. This was because the studies had not compared the risk of children exposed to power-line magnetic fields with that of unexposed children but had compared the risk of children living close to high-current lines with that of children, who, though living near low-current lines, were nonetheless being exposed to power-frequency fields of considerable strength. In a 1991 memorandum that had been published in *Microwave News* and widely circulated in the epidemiological community, Nancy Wertheimer had pointed out that the childhood cancer studies were like studies comparing the risk of lung cancer in people who smoke two and half packs of cigarettes a day with people who smoke two packs. Thus, the twofold increased risk that Neutra disparaged was actually greater than twofold — perhaps as great as sixfold — which should certainly be cause for serious concern on the part

of any public health official. Neutra, however, appeared determined to play down the power-line hazard. He went on
to remind the Rotarians that "everything you do has costs
and risks." He warned them that studies of the power-line
hazard would undoubtedly prove to be very expensive, and
he told them that the controversy surrounding the hazard
would serve to keep trial lawyers and newspaper reporters
"very busy." He said that the members of the California
Consensus Group had agreed that there ought to be more
research into the problem, and that the health department
felt that the general public ought to know the pros and cons
of the controversy "so that somebody who wants to buy an
electric blanket can think about just how risk averse are
they." After suggesting that environmental activists invariably "are trying to get somebody else to fork out the
money," he concluded his remarks by telling the Rotarians
that there would always be problems with new technology,
and that "we're learning how to live in a technological society."

CHAPTER SEVENTEEN

Confrontation

A FTER HIS TALK to the Rotarians, Neutra was driven to the Slater School by DeYoung and a P. G. & E. engineer, who were also attending the subcommittee meeting. There he was informed by Berryman of the inaccuracy of his site plan. She told him that when his error was corrected a large section of the school playground, including the kindergarten sandbox, would fall within the very-high-current zone, and that a considerable portion of Pod A and perhaps part of Pod B could be expected to fall within the ordinary-high-current zone.

After some discussion, Neutra agreed to go outside at the end of the meeting and examine the power lines to see if additional measurements were necessary. He said that any such measurements could be made by the school's maintenance personnel.

Berryman suggested that the Department of Health Services should be taking the measurements. She then asked Neutra if the wiring code could accurately assess the magnetic fields given off by high-voltage transmission lines that supplied power for thousands of homes, and reminded him that London and Peters had called for the code to be refined and improved.

Neutra replied that the code was consistent for both high-current and high-voltage wires, and that "everybody knows this." He asked Berryman, "What alternative would you propose?"

Berryman considered the question unfair. "We got your report on Friday," she told Neutra. "We got this so late. Our questions were given to you on November twenty-second." She went on to point out that she and her colleagues had only had the weekend to review it.

During the rest of the meeting, the confrontation between Berryman and Neutra continued. At one point, Berryman challenged him on the accuracy of his cancer count, pointing out that on page two of the progress report he and Glazer stated that they had confirmed nine cases of invasive cancer among teachers, teacher's aides, and staff members, but that a diagram attached to the end of the report listed ten verified cancer cases. She went on to suggest that there could very well be more than ten malignancies since the diagram indicated that there were two people who did not wish to cooperate with the investigation; three people who denied having cancer; and one person, who was reported to have developed cancer, but could not be located. "How do we know who you're counting or not counting?" she asked.

Neutra's response was to tell her that he might not be able to include several of the confirmed cases of cancer in his study, because they had occurred in people who were not in class pictures and were not on the master list of teachers and teacher's aides that Berryman and Loretta Hutton had sent him back in August.

Berryman reminded Neutra that he had known from the

beginning that some of the cancer cases had developed among nonteaching staff members at the school, and that this was why he had added an estimated six staff members to the total number of school employees in his September progress report, and why he and Glazer had included the same number in their latest report. She went on to say that all the cancer cases should be included in his study, because three of the staff members who had developed cancer had worked within a few feet of Pod A, and because a fourth cancer had developed in someone who had worked only ten to fifteen feet away. "It's not difficult to show you who worked in what room in the administration area, so you can see for yourself," she declared.

Neutra insisted that, since his calculations showed that only a corner of Pod A fell within the ordinary-high-current exposure zone, a truly accurate study would include only those people who had worked in Rooms 2 and 3 of the pod. He went on to describe the wiring-configuration code as "the only Rock of Gibraltar the epidemiologists have" for assessing the association between magnetic-field exposure and the development of cancer.

When Berryman inquired what the Department of Health Services was doing about the children who had attended Slater, Neutra told her that there was no evidence to date of any unusual health problems among them. He said that one could expect about twenty-four cases of invasive cancer to have developed among the thirteen thousand or so students who had attended the school since 1972, and that the number of cases that had been reported so far fell well within that range. He went on to say that the Slater School student population was not large enough to supply an answer to the question of whether power-line magnetic fields posed a health risk to schoolchildren, and that a nationwide study was needed. After complaining that Governor Wilson's veto had hampered the Department of Health Services' program for studying the power-line problem, he said that he had been working "ten days a week since October" as a member of the California Electromagnetic Fields Consensus Group, which had been convened by the Public Utilities Commission. He then claimed that as a result of the group's deliberations he expected that the Department of Health Services would be given additional resources to investigate the association between magnetic-field exposure and leukemia and brain cancer among children attending schools that, like Slater, were situated close to power lines. "My guess is that by this time next year we'll be in business," he declared.

This was the first time that Neutra had mentioned his involvement with the California Consensus Group to Berryman, who said that at this point in the subcommittee meeting, he "seemed to be trying to bring the proceeding to a close by telling us all the wonderful things the state was going to do for us in the future." For her part, Berryman told Neutra that the magnetic-field levels she was now measuring on the Slater School side of the transmission-line corridor were only about half as strong as the fields that P. G. & E.'s engineers had measured in the initial survey they had conducted back in December of 1990, and the levels on the far side of the corridor were approximately double what they had been in the earlier survey. Since she was measuring at the very same sites the P. G. & E. engineers had measured, it seemed clear to her that the one-hundred-and-fifteen-thousand-volt line nearest the school was carrying less current than it had carried in December of 1990, and the two-hundred-and-thirty-thousand-volt line on the far side of the transmission corridor was carrying more current. When Neutra asked DeYoung if the loads on the two lines had been changed, DeYoung replied that the loads were constantly changing, and that this situation made it impossible to get consistent measurements or to predict what the strength of the emitted magnetic fields might be. DeYoung went on to say that there were no records available to show what the loads on the transmission line had been in years past, and therefore no way of knowing what the strength of the magnetic fields may have been.

After the meeting, Berryman, Neutra, DeYoung, and Ray Spena, the Fresno Unified School District supervisor of operations and maintenance, went outside to the kindergarten playground, and Berryman invited Neutra to see for himself that the transmission lines did not follow the curve of Emerson Avenue but turned toward the northwest and passed directly above a corner of the main playground sandbox before running across a city park that was situated next to the Slater School grounds. "Spena and an assistant then measured the distance from the center of the one-hundred-and-fifteen-thousand-volt line to Pod A with a tape measure, and found that the transmission line was about ten feet closer to the school than Dr. Neutra had estimated on the aerial site plan," Berryman recalls. "A few minutes later, DeYoung reminded Dr. Neutra that he had a plane to catch, and Dr. Neutra hurried off."

After Neutra left, Berryman stood in the kindergarten playground trying to take stock of what had happened. "The more I thought about it the more puzzled I became," she said. "Here, seven months after he had taken over the study from Dr. Stallworth, Dr. Neutra was suddenly claiming that the cancers among the staff members who worked in the administrative area next to Pod A didn't count, because they had developed in people who weren't on the list of

teachers and teacher's aides that Loretta Hutton and I had
culled from the class pictures back in August. Why, I won-
dered, hadn't he told us that before? And why had he
counted the cancer cases among staff members in the draft
report he had given us in September? And if the wiring-
configuration code was such an epidemiological Rock of
Gibraltar why hadn't he used it back then? As for the slap-
dash manner in which he had drawn the hazardous zones
on the aerial site plan, I found it hard to believe that a physician and senior health official could be so careless. After
all, we were dealing with a suspected cancer-producing
agent and with the health and safety of children. In this
frame of mind, I remembered that I had forgotten to ask Dr.
Neutra about the most serious flaw in his study — why he
had consigned the cancer cluster at Slater to chance and ab-
solved the transmission line before tracking down and assessing the health histories of the teachers, teachers' aides,
and staff members who no longer worked at the school."

On the following day, February 4th, Berryman and Lynn
Stenson sent a letter to Neutra reminding him of the question that had been raised about the actual magnetic-field
levels to which the teachers and aides working in Pods A
and B had been subjected over the years. "Since all adult
cases of cancer at Slater School occurred among people
working in these areas — the portion of the school closest
to the high-voltage transmission lines along Emerson
Ave. — we believe this question to be of overriding importance," they wrote. "For the same reason we believe it im-
perative that the State Department of Health Services
determine with accuracy, not guesswork, whether or not
there has been an increased incidence of cancer among the
children attending or who have attended Slater School."

*   *   *

Curtis Hurd died on February 27th. Four days later, the *Fresno Bee* published an article by Amy Alexander, which ran under a headline that read "ANOTHER CANCER DEATH RENEWS SLATER WORRIES." Shortly thereafter, Neutra arranged with school-district officials to hold a conference call for the purpose of resolving some of the issues that Berryman and Stenson had raised in their letter of February 4th. The conference call, which took place on the afternoon of March 3rd, began shortly after one o'clock and lasted about an hour and a half. In addition to Neutra, the participants included Dr. Glazer, Anna Phillips, Berryman, and Tanis DeRuosi, who had become principal of the Slater School in August.

"During the call, Dr. Neutra asked me where I wanted the hazardous line to be drawn at the school," Berryman recalls. "I thought it was a highly inappropriate question, but I didn't want to seem antagonistic, so I said that I didn't want a line to be drawn anywhere, but that I did want all of the staff members who had developed cancer to be included in the study, because they had worked so close to the power lines. At that point, Dr. Neutra informed me that if we expected him to include Curtis Hurd, we would have to come up with the names of all of the vice-principals who had worked at the school. He went on to say that if we expected him to include a cook who had developed cancer while working in the kitchen next to Pod A we would have to provide him with the names of all the cooks who had worked there, and that if we expected him to include the custodian who had developed melanoma we would have to furnish him with the names of all the custodians and maintenance personnel who had been employed at Slater. In fact, he informed me that we would now be required to furnish him with the names of all the administrative personnel who had ever worked at Slater. I could scarcely believe my ears. Dr. Neutra was setting our continued participation in gathering data that he and his colleagues at the Department of Health Services should have been gathering as a condition for them to continue to carry out the cancer-cluster study. He made it seem as if they were doing us a favor. When I told him that we had a special concern about the health experience of the children who had attended Slater, he said that a study of childhood cancer at the school would be prohibitively expensive. When I suggested that an advertisement be placed in the *Bee*, asking former Slater students to telephone information about their health to a special number, Dr. Glazer said that this would alarm people unduly. As for our concern about the actual strength of the power-line magnetic fields to which we had been subjected over the years, Dr. Neutra said that he would have to estimate our past exposure on the basis of the average-load data that P. G. & E. had compiled during the late summer and early autumn."

Following the March 3rd conference call, Sandra Craft and Loretta Hutton talked to several senior members of the Slater faculty, who estimated that a total of about twenty people, including vice-principals, cooks, kitchen helpers, and secretaries, had worked in the administrative area over the years. They gave this figure to DeRuosi, and she sent it to Anna Phillips, who passed it on to Neutra. On March 13th, Neutra wrote a letter to Lynn Stenson (with a carbon copy to Berryman) in which he said that he had tried to reach her in person after receiving the letter that she and Berryman had sent him on February 4th, "but our respective schedules never seemed to overlap." Neutra apologized for the delay in sending a written response to the questions she, Conley, and Berryman had sent him back in

November, but claimed that the "essence of my subsequent responses was given verbally at that time over the speaker phone." After requesting that Stenson provide "specific comments" on any aspects of his progress report and appendix with which she and Berryman were dissatisfied, Neutra addressed the issues that Berryman had raised during the March 3rd conference call. He declared that there would be "no positive scientific value" in conducting a full-scale study of cancer among former students at the Slater School, and claimed that "we are pushing hard for a serious study of school EMF exposures within the context of well-designed studies now underway." As for Berryman's suggestion that there had apparently been a major reduction in the amount of current flowing through the one-hundred-and-fifteen-thousand-volt transmission line on Emerson Avenue, he told Stenson that "we have asked P. G. & E. to report on changes in the system and the likely changes in current flow along these lines." The rest of his letter read as follows:

I suggested to Mrs. Berryman today that a third party, trusted by you, might review their records to see if their assessments seemed reasonable. She indicated that no such evidence was likely to change her conviction that fields may have been higher in the past and that these caused the cancer cases seen. She said she was very concerned that the teachers and students may be moved back into Pods A and B and that the lower levels of fields now being seen would be used as a justification for doing this. We will try to make a reasonable effort to address this concern of past exposures, but it is pretty clear that we will be unable to give you the definitive answers you desire.

You are understandably focused on the immediate decisions related to the school. I think our findings will not
give you the kind of information you desire, but a careful
documentation of this or any other similar events which
arise over the next few years, when taken in the context
of ongoing research, may provide a guide for future
policy.

Berryman interpreted Neutra's letter as another attempt
to avoid addressing the issue of why there had been so
much cancer and other disease among employees who had
worked on the side of Slater nearest the transmission lines.
"At no time — either on March 13th, when he claims to
have done so, or earlier — did Dr. Neutra ever suggest to
me that a third party who would be trusted by me or by
anyone else on the subcommittee should review P. G. & E.'s
line-load records and data to see if they seemed reasonable," she said at the time. "I can't imagine why he should
have made such a claim, unless he was trying to divert attention from the fact that by failing to take precise background magnetic-field measurements of his own, and by
accepting the possibly flawed measurements of P. G. & E.,
he had committed the California Department of Health Services to an unseemly close relationship with the utility."
    On Friday, April 3rd, Neutra sent Tanis DeRuosi a copy of
the aerial site plan of the school and asked her to measure
the right-angle distance from the midpoint of the hundred-
and-fifteen-thousand-volt line to Pod A and Pod B. DeRuosi
asked Berryman to help her and on the following Sunday
afternoon the two women met at the Slater School and,
guessing at what constituted a right angle to the transmission line, used a two-hundred-foot-long ball of twine to
measure the distance to the two pods. According to their

calculations, Pod A was a hundred and twenty-six feet from the line, and Pod B was a hundred and forty-six feet from it; both measurements showed the transmission line to be closer to the school than Neutra had indicated in the appendix to his progress report. DeRuosi marked the aerial site plan with the measurements that she and Berryman had made, and drew the path followed by the hundred-and-fifteen-thousand-volt transmission line as it ran past the school and crossed over the southwest corner of the school's main sandbox. On the next day, she mailed the aerial site plan back to Neutra.

On April 28th, Neutra called DeRuosi to ask her if she thought one of the measurements might be inaccurate. When DeRuosi said she didn't know, Neutra told her she might have to do the measurements over. On May 1st, Neutra called Berryman and asked her the same question. "I told him that the measurements Tanis and I had made were the best we could do with the means we had," Berryman recalled. "When I saw Tanis, I told her that, considering the importance of the health issue at Slater, we should not assume any further responsibility for taking measurements that Dr. Neutra ought to be taking and was intending to use in an official state report. Tanis agreed with me, and called Beauregard that afternoon to tell him how we felt about the matter."

On May 15th, DeRuosi showed Berryman the rough draft of a questionnaire that, with Beauregard's approval, she was planning to send to the Slater teachers asking them how they felt about moving back into Pods A and B "in the event Dr. Neutra's summary of findings is inconclusive." Over the weekend, Berryman decided that it would be a mistake for her colleagues to respond to the questionnaire, and on Monday she sent them a note urging them not to fill it out. She also sent a note to Beauregard telling him that it was "inappropriate to be asked to comment on this matter before we have had the opportunity to review the final report from the California Department of Health Services regarding this very important health issue," and reminding him that the letter sent to Olivia Palacio by forty-seven teachers and teacher's aides on May 7th, 1991, informing her that they would not work in Pods A and B after August 10th, was "still in effect."

At a staff meeting on Tuesday, May 19th, Berryman reminded her colleagues that the letter to Palacio had been written out of concern for the health of the teachers and children at Slater, and that it had forced school officials to recognize the potential hazard posed by the transmission lines on Emerson Avenue, and to close down the ten classrooms on that side of the school. She went on to say that she felt that it was very important for them to remain united in their resolve not to return to Pods A and B, and her recommendation was supported by everyone present.

On the following day, Berryman learned that a woman who had been a noontime helper at Slater for four years had just undergone a mastectomy for a malignant breast tumor. By her count, this was the thirteenth case of cancer that had been found in people who had worked on the south side of the school. "The noontime helpers come in for two to three hours a day to supervise the children at lunch and to watch over them when they play outside," she said at the time. "There are usually five such helpers, and until two or three years ago they were usually teacher's aides. When the children eat inside, they eat in the all-purpose room, which is about the same distance from the high-voltage lines as the office used by two secretaries who developed cancer. For at least an hour each day, the helpers pull yard duty — formerly in the main playground, where, until it was closed a year ago, the first-, second-, and third-graders spent their recess time. The sandbox there contains a jungle gym, climbing bars, and chin-up bars, and, since one corner of it lies directly beneath the hundred-and-fifteen-thousand-volt line, the magnetic-field levels are extremely high. In fact, I have measured levels of more than ten milligauss in various portions of this sandbox on several occasions. Strong magnetic fields can also be measured at the basketball court and the four-square area, which have also been closed. All three of these play areas are about the same distance from the transmission lines as the kindergarten playground, where Dr. Neutra and Pamela Long measured a field strength of nine milligauss in September. For this reason, the noontime helpers and the children they were watching over undoubtedly underwent higher exposure to magnetic fields than anyone else who worked in or attended the Slater School. Is it any wonder that the parents joined us last year when we demanded that classrooms and play areas on the south side of the school near the transmission lines be closed? And, considering the fact that over a period of many years these play areas were used almost daily by thousands of children, doesn't it seem reasonable for us to have requested that the Department of Health Services expand its study to include former Slater School pupils?"

CHAPTER EIGHTEEN

Some Conflicts
of Interest

AFTER NEUTRA'S TELEPHONE CALL on May 1st, Berryman did not hear from him until the middle of October, when he left a message on her answering machine that he would soon be sending a draft of his final report on the cancer cluster to Beauregard. Meanwhile, students and teachers at schools elsewhere in California and across the nation were being exposed to power-line magnetic fields even stronger than those which had occasioned the closing of Pods A and B at Slater. Fields of between six and eighteen milligauss had been measured, for example, in kindergarten classrooms on one side of the Brooklyn Elementary School, in the Golden Hill section of San Diego, just a few feet from a high-current distribution line that runs along Fern Street; levels of over twenty milligauss had been measured in classrooms on the east side of the Independence Elementary School, in Bolingbrook, Illinois, which is situated within a hundred feet of a right-of-way containing two high-voltage transmission lines. In May of 1991, Neutra told San Diego Unified

School District officials, who were trying to decide what to do about the situation at the Brooklyn School, that there was no scientific basis for regulating power-frequency fields, and in March of 1992 officials of the Illinois Department of Public Health, who had been asked to investigate the situation at the Independence School, told state legislators that because the data on the power-line hazard were inconclusive, they could offer no advice. Neither Neutra nor his counterparts in Illinois had anything to say about the daily exposure of young children to magnetic fields that were nine to ten times as strong as those which had been associated with a two-to-threefold increase in the incidence of childhood leukemia.

A similarly dismissive attitude was displayed during 1992 by Connecticut health officials. On February 3rd, Susan S. Addiss, Commissioner of the Department of Health Services, and Timothy Keeney, Commissioner of the Department of Environmental Protection, wrote to Senator O'Leary, explaining that they had been too busy with other issues to make as much progress on the electromagnetic-field study as they had expected, and would not be recommending the prudent-avoidance measures he had requested more than a year earlier until autumn. The two officials added that "knowledge gained in the past year indicates that the EMF issue is more complex than we previously thought," and that "therefore, we need to be extraordinarily careful about providing any advice." Two days earlier, Addiss and Keeney had sent the Connecticut legislature an interim report on the task force's effort to come to grips with the power-line hazard. After concluding that "the possibility of adverse health effects associated with EMF has not been demonstrated at this time," they told the lawmakers that "an active campaign of prudent avoidance does not seem appropriate until it is better understood whether or not a hazard exists."

On February 21st, Commissioner Addiss sent O'Leary a copy of the preliminary draft report of the Connecticut Academy of Science and Engineering's findings on the power-line hazard. Because of a delay in enacting legislation to finance the Academy's study, the ten members of an ad hoc committee appointed by the Academy to conduct it had not got down to work until the autumn of 1991. Addiss told O'Leary that she considered the Academy report to be "an excellent contribution to our analysis of this controversial and complex issue," and that nothing in it would lead her to recommend that a statewide study of the power-line hazard should be undertaken. In April, O'Leary sent a copy of the report to Michak, who wrote a story in the *Journal Inquirer* that ran under the headline "STATE'S EMFs STUDY COMES UP INCONCLUSIVE." Michak quoted the report as declaring that "absolute proof of the occurrence of adverse effects of ELF fields at prevailing magnitudes cannot be found in the available evidence, and the same evidence does not permit a judgement that adverse effects could not occur."

The final version of the Academy's report, which came to forty-four pages, was not made available to the public until June 4th. Its authors concluded that the epidemiological studies did not demonstrate that residential or occupational exposure to power-frequency magnetic fields "unequivocally" increased the risk of cancer. They went on to declare that it would be "inappropriate" for health authorities to recommend that the public avoid exposure to the fields. As for the cancer cluster on Meadow Street, they said that it had been "investigated in detail by the Academy but was considered to be uninformative" with regard to any potential health hazard. They went on to say that the Academy had been provided with information indicating that magnetic-field levels in Meadow Street houses "were not unusual," and they agreed with the conclusion of David Brown and his colleagues in the Department of Health Services that no cancer cluster had occurred among the residents of the street.

On June 5th, the *New Haven Register* published a story about the Academy report, which quoted Jan A. J. Stolwijk, the chairman of the ad hoc committee, who is a professor of epidemiology and public health at the Yale University School of Medicine, as saying that there was "no evidence that there are any health effects" caused by exposure to power-frequency magnetic fields. Stolwijk added, "I can't prove that a meteorite won't fall on New Haven tomorrow, but I'm reasonably sure it won't." The article went on to quote Michael B. Bracken, a professor of epidemiology, obstetrics, and gynecology at the Yale University School of Medicine, who had also been a member of the ad hoc committee, and, together with Stolwijk, was said to be the chief author of the Academy report. Bracken dismissed the cancer cluster on Meadow Street out of hand. "Diseases don't fall evenly on every town like snow," he told the *Register*. "You see clusters of any kind of cancer; there are clusters for every kind of disease all over the country. Guilford, it's not even a cluster, because the cancers aren't related. These are individual tragedies." As for the magnetic fields emitted by power lines, Bracken said that "evidence is clearly mounting that either there are no health effects, or they're so small they can't be measured." He went on to insist that concern about the power-line health hazard was unjustified.

Since most of the article in the *Register* was carried by

the Associated Press, the statements of Stolwijk and Bracken found their way into other newspapers in Connecticut. As a result, many Connecticut residents were given the impression that power-frequency magnetic fields had been found to pose no health risk whatsoever. This notion was perpetuated in an editorial that appeared on June 27th in the *Hartford Courant* — the state's largest newspaper — which declared that experts from the Connecticut Academy of Science and Engineering had concluded that electromagnetic fields did not pose a health hazard, and that epidemiological studies had failed to link power-line emissions with any health problems. The *Courant* editorial concluded that since the Academy's experts could not justify prudent avoidance of power-frequency magnetic fields people should do nothing to reduce their exposure to them.

Because of Senator O'Leary's long-standing request that state health officials propose prudent-avoidance measures for the General Assembly to consider, the authors of the Academy report went to some lengths to justify their opposition to such measures. "Members of the Academy committee that performed this study discussed 'prudent avoidance' in a telephone conference with Prof. John Peters, University of Southern California, after he completed his study of childhood leukemia and ELF exposures in the Los Angeles area," they said on page fourteen of their report. "There was concurrence from this discussion, as well as from other information, that it would be inappropriate for health authorities to recommend 'prudent avoidance' to the general public." Since Dr. Peters was a co-author of the highly regarded epidemiological study that had found children living in homes near high-current power lines to be experiencing a two-and-a-half times greater risk of developing leukemia than children living in homes that were not situated near such lines, his opinion about prudent avoidance was bound to carry weight. However, when the passage from page fourteen was read to him a few weeks after the report was issued, he denied that he had said or inferred that prudent avoidance was an inappropriate recommendation. "On the contrary, I consider prudent avoidance to be a very reasonable approach," he declared.

Further evidence that the Academy report contained serious deficiencies could be found on page thirty-four, where its authors turned their attention to the cancer cluster on Meadow Street in the following paragraph:

> The Meadow Street cluster in Guilford was investigated in detail by the Academy but was considered to be uninformative with respect to the general relationship of ELF magnetic fields to health effects. The tragic personal nature of the events on Meadow Street is appreciated. From information provided to the Academy, the magnetic field levels on Meadow Street were somewhat elevated, but with the distribution wiring being across the street from the residences, the levels measured in the houses were not unusual. The diseases reported on Meadow Street were multiple, and there was not a recognizable cluster of any given disease as defined and could be evaluated by procedures of the U.S. Centers for Disease Control. Thus, the Academy concurs with the conclusion reached by the Connecticut Department of Health Services: There is no indication of a recognizable cluster of any type of relevant cancer.

Whether this assessment truly reflected the information that had been given to the Academy committee about the situation on Meadow Street seems doubtful in light of a

letter that was sent to Stolwijk on March 17th by Dr. David
O. Carpenter, who is dean of the School of Public Health of
the State University of New York at Albany, and one of two
physicians who served on the ad hoc committee. In his letter, which contained a number of suggestions for revising
the Academy's initial draft report, Dr. Carpenter wrote as
follows about the Meadow Street cancer cluster:

> I am somewhat surprised, given the charge to the com-
> mittee, that more data on the Meadow Street situation is
> not presented. My recollection from the information pre-
> sented is not consistent with the statement in the first
> paragraph that "from the information provided to the
> Academy, the magnetic field levels in the houses on
> Meadow Street are compatible with levels found else-
> where." I thought we were told that the fields in the street
> and probably the homes were relatively high. I would
> think that we would want to give much more detail here,
> including measured values from both the street and the
> homes, as well as the basis for the reference to the fields
> from "elsewhere." In addition, it would seem to me that
> there should be a much greater discussion of why the
> Meadow Street situation is not a "cluster," which is to say
> that we should indicate what types of cancer were found
> there, with explanation that most of the specific cancers
> are not of the types that have been associated with elec-
> tromagnetic fields in either residential or occupational
> settings.

Considering the fact that two brain cancers, two nonmalignant brain tumors, and a malignant tumor of the eye
that involved a tract of brain tissue had developed among
a handful of people living next to high-current wires and a
substation on Meadow Street, the failure of Stolwijk and the

authors of the Academy report to heed Carpenter's suggestion and discuss the types of cancer that had afflicted the residents there seems puzzling, to say the least. Their assertion that there was "no indication of a recognizable cluster of any type of relevant cancer" seems additionally puzzling, especially in light of their failure to make any mention of the fact that between 1985 and 1991 more than a dozen studies had shown more brain tumors than were to be expected among people exposed to power-frequency magnetic fields at home and at work.

The fact that Dr. Carpenter had allowed his name to appear on the final report that was issued by the Academy in June surprised many observers, who knew that he had been executive secretary of the New York Power Lines Project — a five-year, five-million-dollar research program that had financed the Savitz study which had confirmed Wertheimer and Leeper's original findings — and who had heard him repeatedly state his belief that exposure to power-line fields was associated with the development of cancer. (Indeed, Carpenter had suggested that exposure to power-line emissions might be responsible for up to thirty per cent of all childhood cancer.) However, during the weeks that followed the release of the Academy's report, he apparently regretted having lent it his imprimatur, because, on July 28th — the opening day of a three-day conference on electromagnetic fields that was sponsored by the interagency task force and held at the Ramada Inn and Conference Center, in Meriden, Connecticut — he gave a keynote speech in which he disavowed two of its basic conclusions. Carpenter told his audience — it included Brown, Galbraith, Stolwijk, and Bracken — that the Academy had been wrong to say that there was not enough scientific evidence to support a conclusion that electromagnetic fields could cause cancer.

Pointing to the consistency of results between several childhood-cancer studies and more than two dozen occupational studies, he declared that the weight of the evidence clearly showed that people exposed to power-frequency fields at home and at work were experiencing an increased risk of developing leukemia and brain cancer. He said that recent studies showing increased breast cancer in men who were occupationally exposed to power-frequency fields were particularly worrisome, and he warned that if breast cancer and other reproductive cancers in women were also found to be associated with magnetic-field exposure, the nation would be facing a major public health hazard. He went on to say that contrary to the conclusion drawn in the Academy report Connecticut health authorities should recommend prudent-avoidance measures, such as discarding electric blankets and considering proximity to power lines when buying or renting a house, so that the public could reduce its exposure to power-frequency fields. He added that to do nothing about the problem was unacceptable, because "we are where we were with cigarette smoking twenty-five years ago."

Following his speech, Dr. Carpenter told journalists who interviewed him that one of the chief reasons for the failure of state governments and the federal government to deal effectively with the power-line health hazard was that many scientists who were involved in studying the problem either had ties to or were being financed by the utility industry, and, therefore, had a stake in espousing a do-nothing approach. Later in the day, he emphasized this point by calling upon scientists who were studying the power-line hazard to disclose the sources of their research grants and consultant fees. This prompted Don Michak and another journalist at the conference to look into the possibility that there might be conflicts of interest on the part of some members of the Academy's ad hoc committee. Their inquiries revealed that both Professor Stolwijk and Professor Bracken had established ties with the utility industry before joining the committee.

When questioned about this matter in public on the second day of the conference, Stolwijk acknowledged that he had testified, in a court case in behalf of a state-owned utility in Australia, that exposure to power-line electromagnetic fields posed no health hazard. He claimed that this did not constitute a conflict of interest, however, because he had not charged or been paid a fee for his testimony. Stolwijk also acknowledged that the Academy's investigation of the Meadow Street cancer cluster had consisted chiefly of a briefing conducted by Brown, whose insistence that there was no cancer cluster on Meadow Street had been one of the issues that the ad hoc committee had been charged with independently evaluating. These revelations did not appear to disconcert Stolwijk, who submitted a research proposal to the Department of Health Services in September asking for $374,889 to investigate the association between the development of cancer in Connecticut children and how close they lived to power lines.

As for Bracken, he acknowledged that the epidemiological group he directs at Yale had received a grant from the Electric Power Research Institute to study the effects of electromagnetic fields upon pregnant women. He claimed, however, that the Institute had provided money for only a small part of a much larger program that was being financed by the federal government. Be that as it may, a spokeswoman for EPRI told Michak that under the terms of the grant contract Bracken and his associates would receive two and a half million dollars from EPRI by the end of 1993.

Michak also learned from an official of the Department of Health Services that the department had been aware from the time of the first meeting of the ad hoc committee of the potential conflicts of interest that were implicit in the activities of Stolwijk and the financing of Bracken, but had decided not to make the information public.

CHAPTER NINETEEN

Some Incontrovertible Evidence

UNFORTUNATELY, attitudes and conduct similar to those exhibited by the health officials of Connecticut and California have been the rule rather than the exception in almost every state in which the power-line issue has been raised in recent years. As a result, citizens from one end of the nation to the other have had to fend for themselves in calling for preventive measures to deal with the hazard, and to do so in the face of heavily financed public-relations campaigns mounted by the utilities to convince the public that there was no cause for worry about power-line emissions and, therefore, no need for any preventive measures. All the huffing and puffing by the utilities was designed to obscure the fact that never in history had as much evidence of the carcinogenicity of any agent as had by then accumulated about the cancer-producing potential of power-line magnetic fields been subsequently demonstrated to be invalid and the agent in question found to be benign. Nor was a precedent about to be set. On September 30th of 1992, the wind went out of the utility industry when officials of Sweden's National Board for Industrial and

Technical Development formally announced that they intended henceforth to "act on the assumption that there is a connection between exposure to power frequency magnetic fields and cancer, in particular childhood cancer."

The stunning new Swedish policy was prompted by the results of two major epidemiological studies — one residential and the other occupational — which clearly demonstrated that exposure to electromagnetic fields at home and at work was linked to the development of leukemia. Both investigations had been started in 1987, when health authorities in Sweden, unlike most of their counterparts in the United States, had taken serious note of studies confirming previous findings of excess leukemia and other cancer in children and adults exposed to electromagnetic fields. The residential study was conducted by Anders Ahlbom, a professor of epidemiology at the Institute of Environmental Medicine of the world-renowned Karolinska Institute, in Stockholm, and Maria Feychting, who is a doctoral candidate and research assistant there. With the aid of fellow scientists, Ahlbom and Feychting designed a case-control study to investigate the incidence of cancer among a population of 436,503 people who had lived for at least one year between the beginning of 1960 and the end of 1985 in dwellings that were situated within just under a thousand feet of virtually all of Sweden's ninety-three hundred miles of two-hundred-and-twenty-thousand- or four-hundred-thousand-volt transmission lines. Cancer of all types was studied in children, for whom the one-year residency limit was not applied, while for adults the investigation was limited to leukemia and brain tumors. The cancer cases were identified from records maintained by the Swedish Cancer Registry, and the controls — four for each case of childhood cancer and two for each case of adult cancer — were children and adults without cancer, who were selected at random from the total population living in the designated corridor, and matched to the appropriate cancer cases according to time of diagnosis, age, sex, parish, and power line.

Ahlbom and Feychting assessed the long-term exposure of people living near high-voltage transmission lines by taking spot measurements of field strength in the homes of each case and control, and using them to confirm the accuracy of a computer model that calculated the strength of the fields emitted by each of the lines, according to distance from the line, the wiring configuration of the line, and the current load the line was known to be carrying. They then programmed the computer with records of past current loads that had been maintained for each of the transmission lines by station managers of Vattenfall, a state-owned utility company formerly known as the Swedish State Power Board. The historical load records, which encompassed the entire twenty-six-year period of the study, enabled them to estimate with great accuracy the average annual magnetic-field exposure for each cancer case and its corresponding controls at the time of diagnosis, as well as at intervals of one, five, and ten years before diagnosis.

Upon evaluating their data, Ahlbom and Feychting observed a clear dose–response relationship between increasing magnetic-field exposure and the occurrence of childhood leukemia: children living in dwellings in which they had been exposed to average power-line fields of more than one milligauss experienced twice the risk of developing leukemia as children living in homes in which they had been exposed to fields of less than one milligauss; children exposed to more than two milligauss had almost three times the risk; and children exposed to more than three milligauss had nearly four times the risk. The two scientists found that adults exposed to fields of more than two milligauss experienced a seventy per cent increased risk of developing both acute myeloid leukemia and chronic myeloid leukemia, when compared with adults exposed to less than one milligauss, but this excess was not considered to be statistically significant. Surprisingly, they observed no association in either children or adults between exposure to magnetic fields and the occurrence of brain tumors.

The occupational study, which also employed cases and controls, was conducted by an epidemiologist named Birgitta Floderus and some co-workers in the Department of Neuromedicine at the National Institute of Occupational Health, in Solna, a suburb of Stockholm. It included two hundred and fifty men who had been diagnosed with leukemia; two hundred and sixty-one men diagnosed with brain tumors; and a control group of eleven hundred and twenty-one healthy workers. Workers with exposure to benzene and ionizing radiation — known leukemia-producing agents — were excluded from the study. Floderus and her colleagues assessed the magnetic-field exposure of their subjects by taking more than a thousand prolonged measurements (the mean duration of each was 6.8 hours) in a hundred and sixty-nine job categories. When they analyzed their data, they found that men whose jobs had exposed them to electromagnetic fields that averaged 2.9 milligauss or more developed chronic lymphocytic leukemia three times as often as workers who had been exposed to fields that averaged less than 1.6 milligauss, and that workers whose average exposure was more than 4.1 milligauss experienced an increased risk some four times that of the less-exposed workers. They also found that workers with high magnetic-field exposure developed brain tumors more frequently than workers in the less-exposed group, but this association was not as strong as the association for leukemia.

The fact that both of the Swedish studies demonstrated a clear dose–response relationship between exposure to weak power-frequency magnetic fields and the development of cancer all but demolished the contention of the electric-utility industry in the United States that the excess risk found in previous studies might have been caused by exposure to pesticides, chemicals, or other toxic agents. The fact that both studies showed average magnetic-field exposure over time to be the critical factor in the development of disease blew away the industry's argument that the results of previous studies were suspect, because spot measurements of present-day field strengths could not be correlated with increased cancer risk. And, of course, the fact that both studies were conducted with the financial support and cooperation of the Swedish National Board for Industrial and Technical Development, Swedish governmental health agencies, and the Swedish electric-utility industry contrasted sharply with the lengthy efforts of American utilities to deny the existence of a power-frequency health hazard, and with the pronounced reluctance of state and federal health agencies to acknowledge or deal with the problem.

Shortly after the Swedish findings were announced, Jaak Nöu, the director of the National Board's Department of Electrical Safety, told Louis Slesin of *Microwave News* that Sweden would soon set exposure standards for new homes near power lines, and for all new electrical facilities, and that these standards might require average annual exposures to be in the neighborhood of two milligauss. (In the September/October issue of *Microwave News*, Slesin pointed out that this would be up to a hundred times as strict as the standards for high-voltage line emissions that have been set by New York and Florida — the only two states in the nation to have imposed such limits.) The National Board is also planning to contact all municipalities in Sweden requesting that an inventory be made of schools located near high-voltage transmission lines. In addition, Swedish regulators have declared that they will propose a ban on the construction of houses within three hundred and thirty feet of high-voltage lines. The most difficult problem, of course, will be what to do about homes near existing power lines, because the cost of reducing magnetic-field emissions from those lines will be enormous.

Here in the United States, where the cost of reducing such emissions will also be enormous, the question is whether government health officials and utility officials will follow the example of their Swedish counterparts and acknowledge the existence of the power-line hazard, or whether they will persist in their efforts to deny it. Acknowledging the hazard would require that government and industry undertake to identify where the hazard exists in its most serious and concentrated form — a good place to start might be hundreds of schools and day-care centers that have been built perilously close to high-voltage or high-current power lines — and then set about to remedy the hazard by rerouting such lines, or burying them in a manner that will drastically reduce their magnetic-field emissions. (The technology for doing so was developed and tested by the utility industry some time ago.) Given the record of the government and the industry in dealing with the power-line problem thus far, however, no one should expect that any of this will occur soon, if at all, unless a concerned and determined citizenry forces action.

As it happens, determined and prescient citizens have been confronting the power-line health issue across the nation for a number of years. Early and effective opposition to high-voltage transmission lines was organized during 1987 and 1988 by Citizens Against Overhead Power Lines Inc., a group of homeowners in the Highline section of South Seattle, in Washington State, who prevented Seattle City Light from constructing a pair of two-hundred-and-thirty-thousand-volt lines that would have put strong magnetic fields in dwellings situated on either side of State Route 509, between South 98th and South 136th Streets. In 1990, opposition mounted by Residents Against Giant Electric (RAGE), a group of citizens in Monmouth County, New Jersey, who were concerned about the cancer hazard posed by power-line emissions, forced the Jersey Central Power & Light Company to abandon its plan to construct a pair of two-hundred-and-thirty-thousand-volt transmission lines through the towns of Red Bank, Middletown, Holmdel, Hazlet, and Aberdeen. That same year, the efforts of a group called Rhode Islanders for Safe Power persuaded the town council of East Greenwich, Rhode Island, to enact a three-year moratorium on the construction of power lines carrying more than sixty thousand volts in that town, and to close the East Greenwich High School's soccer playing field, where magnetic fields of more than four milligauss had been found to be coming from a nearby hundred-and-fifteen-thousand-volt line that runs through East Greenwich on its way from Warwick to North Kingstown. The East Greenwich moratorium and similar bans that were enacted in the neighboring towns of Foster and Coventry were credited with persuading the Narragansett Electric

Company to drop its plans for building a forty-four-mile-long, three-hundred-and-forty-five-thousand-volt transmission line from Burrilville to Warwick. However, the utility is proceeding with plans to construct an additional hundred-and-fifteen-thousand-volt line in the right-of-way from Warwick to North Kingstown, and officials of the Rhode Island Department of Health have so far refused to investigate a suspicious cancer cluster that exists among residents of East Greenwich whose homes abut this corridor.

During 1990 and 1991, residents of the historic Old Town section of Alexandria, Virginia, who had discovered that high-current distribution wires carrying power to the city's business district were creating magnetic fields of up to forty milligauss in many homes, persuaded city officials to negotiate a franchise-and-operating agreement with the Virginia Electric & Power Company, calling for the city and the utility to share the cost of burying the offending lines, and of redesigning power distribution in a thirty-six-block downtown area. Meanwhile, in New York City, pressure exerted by several citizen-action groups has helped persuade the members of the Manhattan Borough Board to pass a resolution calling upon the City Council to impose a moratorium on the construction and expansion of electric substations by Consolidated Edison and New York City's Transit Authority. There can be little doubt that such measures are needed, because many of the three-hundred-odd substations that are scattered throughout the city abut apartment and office buildings, and some have even been built next to schools. (The Transit Authority has released a list showing the locations of the two hundred and seven substations it operates, but Con Edison has refused to divulge the locations of some of its substations on the grounds that to do so might make them targets for terrorist attack.) Moreover, since few if any of the high-voltage and high-current wires that lead into and out of these substations have been buried in a way that mitigates their emissions, magnetic fields of extraordinary strength can be found on sidewalks from one end of the city to the other. Indeed, a researcher wandering at random through Manhattan recently measured a magnetic-field level of eighty milligauss at the southwest corner of Lexington Avenue and Seventy-eighth Street; a level of ninety-five milligauss a few doorways east of the Harvard Club, on Forty-fourth Street; a level of eighteen milligauss in front of a store on the west side of Sixth Avenue, near Fourteenth Street; and levels higher than ten milligauss at more than a dozen other places along the way.

In 1991, a thousand landowners, business owners, and local governments in Michigan filed a class-action petition before the Michigan Public Service Commission to block a proposal by the Consumers Power Company to construct a three-hundred-and-forty-five-thousand-volt transmission line that would connect power grids in Michigan and Indiana, on the ground that it was unnecessary and would pose a hazard to people and livestock along its right-of-way. In the autumn of 1992, a Commission administrative law judge recommended that the line be built, even though Commission staff members had issued a report finding that Consumers Power did not need the line. On January 28th, 1993, the Commission voted two-to-one that the proposed line would benefit Michigan residents and would not pose a health hazard. However, on February 19th, 1993, a Calhoun County circuit judge ruled that the utility had not proved that the transmission line was necessary, and that Consumers Power could not condemn privately owned land for the project. Environmentalists around the nation said that the ruling would encourage grass roots efforts to halt the construction of new transmission lines in residential areas.

In Pennsylvania, nearly nine thousand citizens have filed protests against a plan by the Duquesne Light Company, of Pittsburgh, and General Public Utilities, of Parsippany, New Jersey — the owner of the nuclear plant at Three Mile Island — to build a two-hundred-and-sixty-eight-mile-long five-hundred-thousand-volt transmission line across fourteen counties, in order to bring power from western Pennsylvania to the eastern part of the state and to New Jersey.

Particularly intense concern about the power-frequency hazard has been raised during the past two years in Illinois, where opposition to the activities and policies of the Commonwealth Edison Company, the largest utility in the state, has been mounted by the members of several community and grass roots organizations — among them a group called Mothers Against Commonwealth Edison (MACE) — who have battled fiercely to prevent the company from constructing new power lines and electric substations in residential areas and near schools. In 1991, Mayor Richard M. Daley, of Chicago, denied Commonwealth Edison permission to erect a three-hundred-and-forty-five-thousand-volt line along Metropolitan Rail's train-track corridor between Twenty-third Street and Sixty-first Street, on the ground that the electromagnetic fields given off by the line might endanger the health of people who live or work in its vicinity. (Officials of the Chicago Housing Authority, which operates apartment buildings occupied by some forty thousand people living within a half block of the train tracks, were outraged at not being consulted by the utility about its plans to build the line.) Restrictions have been placed upon the size and location of electric substations in

Wheaton, Evanston, and Antioch, and efforts are being made to limit magnetic-field emissions from Commonwealth Edison's power lines in dozens of other cities and towns across the state. In the Village of Lincolnwood, a suburb with twelve thousand inhabitants that is located just north of Chicago, a study showing that magnetic fields of more than seven and a half milligauss exist a hundred feet from a Commonwealth Edison right-of-way containing a high-voltage transmission line and several high-current distribution lines has prompted Village officials to demand that the utility take steps to reduce the magnetic-field emissions to a maximum level of a milligauss and a half at the setback line of any building in the right-of-way. In Aurora, a city of some eighty-five thousand inhabitants located about forty miles southwest of Chicago, members of a group called Parents Against High Voltage Education have persuaded school-district officials to move the site of a new school away from the vicinity of some high-voltage transmission lines.

Elsewhere in Illinois, the situation existing at the Independence Elementary School in Bolingbrook — a city of forty thousand inhabitants located about thirty miles southwest of Chicago — illustrates how differently citizens in different parts of the nation may react to a similar situation. During the spring of 1991, while teachers and parents of children attending the Slater Elementary School were demanding that the Fresno Board of Education close down the south side of the school because of the hazard posed by the high-voltage transmission lines on Emerson Avenue, some parents of children attending Independence School, which is within seventy to a hundred feet of a Commonwealth Edison right-of-way containing a three-hundred-forty-five-thousand-volt transmission line and a hundred-and-thirty-eight-thousand-volt line, voiced similar concerns and asked their school-district officials to investigate the problem. At the request of school officials, engineers from Commonwealth Edison came to Independence on April 30th, and took magnetic-field readings throughout the building. As might be expected, the highest levels were found in second-, fourth-, and fifth-grade classrooms that are located on the east side of the building, nearest the power lines, where magnetic-field strengths ranging from two and a half to eight milligauss were measured.

On May 17th, Edward P. Carli, the principal of Independence, sent a letter to parents telling them that students in one classroom had been relocated "as a 'prudent avoidance' measure in order to ensure 'peace of mind' until the end of the school year." He did not mention that this was a fourth-grade classroom in the northeast corner of the school, where the Com Ed engineers had recorded the reading of eight milligauss, or that the relocation had been undertaken at the insistence of the fourth-grade teacher, who had developed cancer after working at Independence since 1976, when the school had opened. He did, however, tell the parents that scientists had conducted many "conflicting" studies of the power-line cancer hazard, and that no federal or state agency had concluded that there was a danger from overhead transmission lines. He went on to say that "since the scientific community cannot be conclusive with its results, it is very difficult to determine what is right and what is wrong." Carli ended his letter by assuring the parents that the power-line hazard "is and will remain a top priority at Independence School," and that they would be kept informed of developments.

Four days after Carli sent his letter, engineers from the electromagnetic-field-measurement program of the University of Illinois' Institute of Electrical and Electronics Engineers came to Independence and recorded a magnetic field of eighteen milligauss in the fourth-grade classroom that had been evacuated. The readings they took throughout the school were two to three times as high as those that had been taken by the engineers from Commonwealth Edison. The University of Illinois technicians found levels higher than five milligauss in thirteen of the school's twenty-four classrooms, and fields of nearly two milligauss in the kindergarten which is located in the corner of the building farthest from the transmission lines. On the same day, these technicians measured magnetic-field levels at the Jonas Salk Elementary School, which has the same layouts and floor plan as Independence but is located several miles away in the eastern section of Bolingbrook and is not near any high-voltage or high-current power lines. The ambient fields at Jonas Salk ranged between .1 and .4 milligauss. On May 24th, Carli sent a second letter to school parents, informing them in general about the two sets of magnetic-field measurements that had been taken at Independence. He said nothing about the measurements that had been taken at Jonas Salk, but he assured the parents that school district authorities would continue to monitor the situation. On May 28th, engineers from the University of Illinois returned to Independence and measured even higher magnetic-field levels than before, including a level of twenty milligauss in the classroom that had been evacuated. That evening, the Valley View Board of Education, which has jurisdiction over Independence, held a meeting at the school to review conditions there with concerned parents. At the meeting, an environmental consultant for the Board told the parents that there was no conclusive evidence that power-line magnetic fields caused cancer, and that "the jury is still out" on the matter. Two days later, William Harm, a reporter for the *Chicago Tribune*, wrote that school officials had told him that magnetic-field levels at Independence were "low" and posed "no threat to children or adults in the building."

At the end of June, a task force that included Carli, two local physicians, a representative from Commonwealth Edison, and several residents of the school district asked the Illinois Department of Public Health to look into the hazard posed by the power lines. During the summer, members of a group called Parents for a Safe Environment (the same name that parents at the Slater School had chosen for their organization) circulated a petition asking the Board of Education to continue investigating the problem, and got the signatures of more than a hundred parents. They also contacted the local representative of the American Federation of Teachers — the union that represents the teachers at Independence — who told them that until there was conclusive proof that a power-line health problem existed, she did not intend to "scare our members." (A number of teachers at the school had already voiced concern that publicity over the power-line hazard might cost them their jobs.) On August 29th, members of Parents for a Safe Environment described the findings of the childhood cancer studies that had been compiled in the E.P.A. report at a meeting of the Independence Home-School Organization — the local version of the PTA — but they were unable to persuade a single parent or teacher in the group that the magnetic fields measured at the school posed any health risk. By this time, the efforts of Parents for a Safe Environment to alert their fellow citizens in Bolingbrook about the cancer hazard posed by the power lines had been severely criticized by local homeowners, who feared that continued publicity about the problem might discourage home sales and lower property values.

Meanwhile, the fourth-grade teacher with cancer, who had insisted that her classroom be relocated, had transferred to another school, and at the beginning of the 1991 school year, in September, fourth-graders at Independence were once again assigned to classrooms next to the transmission-line corridor, where magnetic fields of twenty milligauss had been measured. That autumn, members of Parents for a Safe Environment gave up their efforts to resolve the power-line issue, and since then there has been little formal discussion of the problem by anyone at the school.

In March of 1992, officials of the Illinois Department of Public Health told state legislators that because the data on the power-line hazard were inconclusive, they could offer no advice on how to deal with the problem. As it turned out, the health authorities had made no attempt to evaluate the health experience of the teachers and students at Independence, but had simply examined the residential locations of Bolingbrook residents who had developed leukemia and brain tumors, and determined that no cases of these types of cancer had occurred among people living within eight hundred and fifty feet of the transmission lines. The adequacy of this investigation was called into question during the winter and spring of 1993, when some of the teachers at Independence learned that there appeared to be an unusual incidence of cancer and other tumors among those of their colleagues who had worked on the east side of the school nearest the transmission lines. (There are about forty teachers and teacher's aides at Independence at any one time, and approximately one hundred teachers have worked at the school since it opened in 1976.) Indeed, an informal survey soon showed that in recent years at least seven teachers who worked on the east side of the school had been stricken with cancer. Two of the teachers had developed brain cancer; two had developed cervical cancer; one had developed breast cancer; one had been afflicted with colon cancer; and a teacher who never smoked had died of lung cancer. In addition, two teachers with classrooms on the east side of the school were suffering from adrenal gland problems, which have been linked to magnetic-field exposure, and several teachers working there and elsewhere in the school developed tumors or cancer of the uterus. Moreover, a child who attended Independence had developed and died of brain cancer, and a teacher who had worked for sixteen years on the side of the school near the power lines had given birth to a child who died of an immune-system disease.

CHAPTER TWENTY

How Many More Cancers Will It Take?

A S SOME of the homeowners in Bolingbrook had feared, significant devaluation of residential property situated close to high-voltage and high-current lines has already occurred in various sections of the nation where the power-line frequency hazard has become known, and even greater devaluation appears to be in the offing as word of the latest Swedish findings and other information about the problem is disseminated. Together with the health hazard itself, the devaluation is resulting in widespread concern and litigation. Over the past several years, numerous lawsuits have been brought against utilities in states from Maine to California and from Washington to Florida, and hundreds, perhaps thousands, more are clearly in prospect. As early as 1985, a jury in Houston, Texas, found "clear and convincing evidence" of potential power-line health hazards, and awarded damages to a school district that had brought suit against the Houston Lighting & Power Company for installing a high-voltage transmission line on school property. (The utility subsequently removed the line at a cost of more than eight and a half million dollars.) In 1988, the residents of nearly a hundred homes next to a power-line right-of-way in Wellington, Florida, brought a class-action lawsuit against the Florida Power & Light Company in an effort to obtain damages for reduced property values, and to prevent the utility from adding a pair of two-hundred-and-thirty-thousand-volt lines in the right-of-way. After a three-year legal battle, the plaintiffs and the utility reached an out-of-court settlement in which the utility agreed to postpone construction of the additional lines until further scientific evidence clarifies the nature and extent of the hazard posed by power-frequency magnetic fields. (The recent Swedish studies have surely done that.) In nearby Boca Raton, parents of children attending the Sandpiper Shores Elementary School, which is situated close to some high-voltage lines owned by Florida Power & Light, have agreed to settle a three-year-old lawsuit against the Palm Beach County School Board in return for the right to transfer their children to other schools in the county. (A Palm Beach County circuit court judge had previously ordered that a major portion of the school playground be placed off-limits to the one thousand children who attend Sandpiper Shores, because of the potential hazard posed by magnetic fields given off by the lines.) Elsewhere in the state, attorneys for Hillsborough County, where the Florida Power Corporation has proposed to construct a five-hundred-thousand-volt line, sued the Florida Department of Environmental Regulation for promulgating a standard that allows magnetic-field levels of up to two hundred and fifty milligauss to exist at the edge of some transmission-line rights-of-way. County officials have asked that a three-milligauss limit be established at the edge of rights-of-way for power lines carrying up to two hundred and thirty thousand volts.

Especially worrisome from the utility industry's point of view are several landmark personal-injury lawsuits that have been filed in the last two years. Amng them are the actions brought against Northeast Utilities and Connecticut Light & Power in behalf of the brain-tumor victims on Meadow Street, and a lawsuit brought against the Georgia Power Company, of Atlanta, and the Oglethorpe Power Company, of Tucker, Georgia, in behalf of a thirty-five-year-old woman who is alleged to have developed non-Hodgkin's lymphoma as a result of exposure to magnetic fields given off by high-voltage lines owned by these companies. The lawsuit brought against San Diego Gas & Elctric in behalf of a four-year-old child who was alleged to have developed kidney cancer after being exposed to magnetic fields of between four and a half and twenty milligauss as a result of living in a house next to one of the utility's high-current distribution lines came to trial in April of 1993. After a four-week proceeding, the jury in the case found that the utility, which spent more than $2 million defending itself, had not been negligent for not warning its customers about the potential hazard of power-line magnetic fields at the beginning of 1987, when the child was conceived, and that the high-current distribution line next to the house did not constitute a nuisance. This outcome will undoubtedly encourage utilities that have been named as defendants in other cases to seek jury trials rather than settle out of court. Either way, the industry as a whole may be facing a legal nightmare, because a major settlement or a verdict in favor of the plaintiffs in any power-line case can be expected to trigger an avalanche of similar lawsuits.

Meanwhile, fearful of saying or doing anything that might acknowledge liability, the utilities have been trying to put the best possible face on the medical and scientific evidence showing that power-line magnetic fields are associated with the development of cancer. Typical in this regard was an effort made by Northeast Utilities and Connecticut Light & Power, which sent out a "Consumer News" flyer to thousands of customers in October of 1992, just after the results of the Swedish studies were announced. The flyer told its readers that "scientific research continues to be generally inconclusive on health risks from exposure to EMF," and that "a number of statistical studies have found small links between indirect measures of magnetic-field exposures and the occurrence of some forms of cancer." The flyer promised to keep the utilities' customers "informed of research progress," and, in the event these customers wished to have more detailed information, it offered to send them the discredited review that had been conducted by the Connecticut Academy of Science and Engineering.

In an attempt to involve other culprits in the disease hazard associated with power-line magnetic fields, the utility industry has issued a barrage of brochures suggesting that household appliances, such as hair dryers, toasters, electric ranges, and vacuum cleaners give off strong magnetic fields, and may provide a major source of exposure. What the authors of these industry brochures invariably fail to point out is that the magnetic fields emanating from most household appliances fall off sharply within a few inches of the appliance, and that since appliances tend to be used intermittently, they are not likely to be a source of chronic, long-term exposure. Indeed, Canadian researchers who measured magnetic fields given off by ninety-eight different appliances have reported that although the fields might provide intense exposure to some of the body's extremities, they were not a significant source of whole-body exposure.

Of far greater concern than appliances are the magnetic fields given off by faulty household wiring; by high-current conductors concealed in the walls, ceilings, and floors of commercial office buildings and other large structures; and by high-voltage transformers that can be found in almost any large building. Magnetic fields of well over a hundred milligauss were recently discovered, for example, in a state office building in Madison, Wisconsin, and in a bank office in Manhattan. In both instances, the fields were found to be coming from transformers on the floor below, and were detected only because they caused serious malfunctioning in video-display terminals on the floor above. As for the VDTs, of which some fifty million are estimated to be in use in the United States, they operate by generating extra-low-frequency magnetic fields, and thus provide an important additional source of long-term, chronic exposure. Pulsed magnetic fields of between one and five milligauss can be measured routinely at a distance of twelve inches from the screens of many VDTs, and fields two to three times that strong can be measured at the same distance from their sides, backs, and tops.

In spite of several studies showing that women who work with VDTs have a higher rate of miscarriage than other women, and other studies showing that VDT emissions can damage the fetuses of test animals, government health officials in the United States have failed to limit the magnetic fields that these devices are allowed to give off. Early in 1992, however, researchers at the Institute of Occupational Health, in Helsinki, found that women working with VDTs that exposed them to magnetic fields averaging about three milligauss suffered miscarriages at a rate close to three and a half times that of women using VDTs that exposed them to fields that averaged about one milligauss. The fact that this dose–response relationship is similar to the one established by the recent Swedish studies for exposure to power-line magnetic fields and the development of leukemia is serious cause for concern. Additionally worrisome is a recent study conducted by researchers at the University of Adelaide's Department of Community Medicine, who found that women working with computer monitors are developing primary brain tumors at nearly five times the expected rate. Considering the elevated rates of breast cancer in men with occupational exposure to electromagnetic fields, the sharp increase in recent years in the incidence of breast cancer among women, and the fact that women make up a majority of the work force using VDTs, it seems reasonable to wonder why no government health agency has undertaken an epidemiological study to determine whether women who work with VDTs are developing breast cancer more readily than other women.

The fact that more studies will be needed to define the full dimensions of the magnetic-field hazard should not, however, be used by the nation's public health officials as an excuse for further delaying measures to reduce exposure to power-line emissions. The public whose health these officials are supposed to protect should demand that the power-line hazard be addressed promptly and straightforwardly, and it should be prepared to hold high-ranking health officials and the politicians who appoint them accountable if they fail to do so. The public should understand that by extending the presumption of benignity to power-frequency magnetic fields in the absence of absolute proof of their carcinogenicity, public-health officials have already helped to perpetuate a situation in which thousands of people are at risk of developing malignant disease that could have been avoided. Indeed, in testimony submitted to the California Public Utilities Commission on September

15th, two weeks before the results of the Swedish studies were announced, Neutra himself declared that "if the epidemiological results with regard to childhood cancer and occupational cancer were to be confirmed as real we would be talking about hundreds of preventable cases of cancer in the state each year."

A month or so after making this statement, Neutra finished a thirty-two-page draft of a report entitled "Exposures at the Slater School" and sent it to Donald Beauregard, who passed it along to Patricia Berryman and other members of the Slater Electromagnetic Field Study Subcommittee. In a cover letter dated October 27th, Beauregard informed the subcommittee members that Neutra wanted them to study the draft and mark it with any comments they might have, so that he could review the comments in preparing a summary of findings for a subcommittee meeting that would be held in late November. Most of the draft report was written in question-and-answer form, and Neutra used a considerable portion of it to review the magnetic-field measurements that engineers from Pacific Gas & Electric had taken at the school on various occasions, and to describe the magnetic-field levels that were predicted to occur there on the basis of a computer model that had been developed by P. G. & E. Relying on the utility's measurements and predictions, he declared that during most of the year the magnetic fields at Slater — including the fields in classrooms closest to the transmission lines — would resemble those which could be found in typical residential dwellings. In the margin next to this assertion, Berryman inquired whether P. G. & E. could be considered to have a conflict of interest in the matter, and whether typical residential dwellings were situated next to high-voltage transmission lines.

In the final question of the report, Neutra asked whether

the magnetic-field exposure conditions at Slater could have
caused a cancer cluster. He began and ended a discursive
four-and-a-half-page reply by saying that he could not give
a definitive yes or no answer, but he went to some length
to suggest that there was probably no association between
the fields given off by the transmission lines on Emerson
Avenue and the cancers that had developed among the
teachers and staff members who had worked on that side
of the school.

Berryman was disappointed to find that there was neither confirmation nor denial of her suspicion that emissions from the power lines on Emerson Avenue might be
connected with the cancers that had occurred among her
colleagues. On November 5th this suspicion was reinforced
by the distressing news that still another of her fellow
teachers had developed cancer. The latest victim — the
fourteenth, by her count — was a fourth-grade teacher in
her early fifties, who had worked in Pod B for at least two
years, and had just been operated on for a malignant tumor
of the colon.

On the next day, Beauregard sent Berryman some additional material from Neutra with a cover letter urging that
it also be marked with comments. Among the new documents was a thirteen-page report on the cancer cases at Slater Elementary that had been written by Neutra and Glazer,
who said that they had confirmed eleven cases among past
and present teachers and employees. The eleven confirmed
cases included three breast cancers, two uterine cancers,
two ovarian cancers, two melanomas, the brain cancer that
had killed Katie Alexander, and the colon cancer that had
killed Curtis Hurd. The list was incomplete, because
it did not include two cases of cancer in employees
who had declined to participate in the study, or the colon

cancer that had been diagnosed in the Pod B teacher who had just undergone surgery. The difference between eleven and fourteen cases was extremely important, because, after estimating once again that a thousand out of California's eight thousand schools might be situated next to power lines, Neutra and Glazer went on to declare that eleven cases of cancer could be expected to occur by chance alone among the teachers and staff members of no fewer than fifty-six of these thousand schools. However, according to a Poisson distribution graph that the two health officials had attached to their report, only five and a half of the thousand schools could be expected to have fourteen cases of cancer, and less than one school out of the thousand could be expected to have sixteen cases.

The fact that only two more cancer cases would place the Slater School in a unique category among California schools was not lost on Berryman, who had suspected since January that the incidence of cancer among her colleagues was being underreported. At that time, she and Doris Buffo had telephoned fifteen or twenty former Slater teachers in the Fresno area, and learned that none of them had been approached by the Department of Health Services; this indicated that Neutra and Glazer, instead of going out and conducting a thorough survey of the cancer rate among Slater teachers and staff members, including the eighty-five or so who no longer worked at the school, had merely compiled a list of the cancer cases that had been reported to them, and had thus failed to consider the incidence of cancer in well over half of the population they were supposed to be studying. In a margin on the first page of their report, Berryman wrote the following sentence: "The credibility of your findings is seriously weakened by your failure to determine whether the teachers, teachers' aides, and other staff members who no longer work at Slater have developed cancer, and whether there has been any increased incidence of cancer among the students who have attended the school." She went on to note the colon cancer that had just been reported in the Pod B teacher, and the fact that the number of cancer cases had doubled since the two health officials had begun their investigation, a year and a half ago — all of the cancers having occurred among people working on the side of the school nearest the transmission lines. Then she posed a question that, unless remedial action is taken, seems likely to be asked again and again in the coming years by people living on Meadow Streets across America, and working in schools next to power lines: "How many more cancers will it take?"

CHAPTER TWENTY-ONE

Some
Revising of Views

THE EXTENT to which California utilities were prepared to deny that any health hazard could be posed by
power-line emissions, and to rely on the State Department of Health Services to keep the whole matter in the
realm of conjecture and further study became apparent
during the week of December 7th, when the utilities and
other members of the California Consensus Group testified
in evidentiary hearings that were held in San Francisco before administrative law judge Michael J. Galvin, of the Public Utilities Commission. What seemed to be foremost on
the minds of utility officials was revealed during the opening minutes of the hearings by E. Gregory Barnes, an attorney for San Diego Gas & Electric, who had previously
sought to equate the involuntary risk of being subjected to
power-line emissions with the voluntary risk of living in an
earthquake zone. Barnes asked Judge Galvin to require that
any lawyers or consultants for parties involved in power-
line litigation identify themselves and their affiliations for
the record. When the judge asked Barnes why he was making such a request, Barnes replied that all of the utilities

present had either been sued or threatened with lawsuits, and that "anything said in here might be offered in other contexts as an admission." After hearing objections to Barnes's motion, Galvin denied it, telling Barnes, "This is a public proceeding and it shall remain as such."

The first witness at the hearings was Jack Sahl, the research scientist for Southern California Edison, who testified that evidence linking exposure to power-line electromagnetic fields and adverse health effects was too inconclusive to warrant any preventive action, and that no weight should be assigned to the recent Swedish studies, because they had not been officially published. (Soon thereafter, the Swedish childhood cancer study was accepted for publication by the *American Journal of Epidemiology,* and was scheduled for publication the following summer.) Sahl said that Southern California Edison had recently completed a study of its work force, which concluded that there was no increased mortality from leukemia, brain cancer, and lymphoma among workers exposed to high-level electromagnetic fields, as compared with workers exposed to lower levels. As it happened, Sahl was the lead author of this study, which was subsequently published in the March 1993 issue of *Epidemiology,* and was conducted with the help of Sander Greenland, an epidemiologist at the School of Public Health of the University of California at Los Angeles, and Michael Kelsh, who worked at U.C.L.A. and at a company called EcoAnalysis.

On March 15th, Sahl told the *Los Angeles Times* that the study "gives me more confidence that there isn't a large problem with EMF and it provides support for the idea that there is no problem with EMF in the workplace." Sahl's optimism was not shared in Sweden, where, several months earlier, a worker's compensation board had recognized brain cancer in an electrician as a work-related injury. Moreover, the accuracy of his and Greenland's findings was questioned by Dr. Samuel Milham, the epidemiologist who had first reported a link between occupational exposure to electromagnetic fields and an increased risk of cancer. Milham pointed out that Sahl and Greenland, who had examined the health records of 36,221 Southern California Edison employees who had worked for the utility for at least a year between 1960 and 1988, should have instituted a cut-off date well before 1988 to allow for the fact that young, recently hired workers could not be expected to develop, let alone die of, cancer within just a few years. Milham also pointed out that rather than publish the results of an internal company comparison — i.e., cancer deaths among heavily exposed workers compared with cancer deaths among less exposed workers — Sahl and Greenland should have published deaths among the California workers compared with expected deaths based on United States or California age-specific mortality rates. Indeed, Milham said that his preliminary analysis of Sahl and Greenland's data showed that deaths from leukemia, brain cancer, and lymphoma made up a larger percentage of all cancer deaths among the Southern California Edison workers than is usually seen in similar cohorts.

When asked at the Public Utilities Commission hearings what implication the decision of the Swedish government to assume that exposure to power-frequency magnetic fields was associated with the development of cancer had for California, Sahl replied that it seemed "somewhat weird that we would somehow pervert our processes" because of action the Swedes were planning to take. He went on to say that "the Swedish have no jurisdiction here in California." He also said that the question of how much scientific value to place on the Swedish studies ought to be determined by the California Department of Health Services, which should be placed in charge of a California research program into the power-line hazard, and given responsibility for educating the public about it.

Almost six and a half years earlier, in July of 1986, Sahl had been the project manager and a co-author of a ninety-four-page report entitled "A Critical Review of the Scientific Literature on Low-Frequency Electric and Magnetic Fields: Assessment of Possible Effects on Human Health and Recommendations for Research." The report stated that the possibility of occupational or community exposure to sixty-hertz electromagnetic fields being associated with cancer, especially leukemia, "cannot be disregarded and should be seriously considered given the present data." The report also said, "The epidemiological findings on cancer are of particular importance because of their direct relevance to human health under exposure conditions commonly found in the environment. The small magnitude of the effect does not diminish the importance of defining a cause-and-effect relationship if one truly exists" because "exposures are life-long and widespread," and "even low *rates* of incidence may translate into large *numbers* of affected persons."

The second and third days of the hearings were largely given over to testimony by Neutra, who complained that the Department of Health Services did not have the resources to provide a timely response to cancer-cluster investigations, such as those it had been asked to conduct at the Montecito Union and Slater elementary schools, and that this had not only caused "considerable anxiety" on the part of the people involved, but also created a situation in which the utilities involved might be forced to make unnecessary and costly changes in their operations. When asked about

the written testimony he had submitted, which acknowledged that if the childhood and occupational cancer studies
were to be confirmed as real, "we would be talking about
hundreds of preventable cases of cancer in the state each
year," Neutra acknowledged that if the power-line cancer
hazard for children proved to be real, it would rank higher
on the list of environmental exposures than chemical exposures.

Later in his testimony, Neutra repeated his earlier assertion that one thousand out of California's eight thousand
schools were probably located near power lines, and that at
least fifty of them could have cancer clusters "just by chance
alone." According to Neutra, this meant that "we'd have to
get up to fifty such schools before . . . we would know for
sure that things were out of line." As for the Swedish childhood cancer study, Neutra acknowledged that its authors
had been able to accurately assess historical exposure to
the transmission-line magnetic fields. However, like Sahl,
he pointed out that the number of cancer cases was small,
and suggested that because of this the Swedish study was
not extensive enough. When asked if this might not also be
the case with studies yet to be conducted under the proposed California research program, he admitted that it was
possible. He went on to claim, however, that within four to
five years the program would initiate and complete a number of significant studies that would provide answers to the
question of whether power lines posed a health problem.
Two and a half years earlier, he had assured parents at the
Montecito Union School that it would take two more years
for such answers to be forthcoming.

When asked what impact the Swedish government's decision to recognize the existence of a link between magnetic-field exposure and cancer would have on the

California Department of Health Services, Neutra replied that he took the Swedish decision to be "more a statement of risk management philosophy than a statement of scientific evaluation." When asked if he believed that the Swedish studies demonstrated a causal link between electromagnetic-field exposure and the development of cancer, he replied that "at this particular point, I don't think that they do."

Later in his testimony, an attorney for San Diego Gas & Electric showed Neutra an article that had been written by Dr. A. A. Afifi, who is dean of the School of Public Health at the University of California at Los Angeles, and a member of the EPRI scientific advisory committee, as well as a former member of a Department of Health Services' advisory committee. The article, which appeared on the op-ed page of the *San Diego Union-Tribune* on December 2nd, had run under a headline that read "SCARE STORIES GENERATE GROUNDLESS EMF FEARS," and in it Afifi asserted that there was no public health reason for rerouting or burying power lines. "Any move now to reduce EMF exposures suggests a perception of risk that is not justified by the available biomedical evidence," Afifi declared. He also said that it would take more research over "another few years" before public health officials and scientists would know enough to formulate public policy regarding exposure to power-line emissions.

When the S. D. G & E. attorney asked Neutra if he considered Afifi to be a fair and objective public health expert, Neutra replied that he did not believe Afifi to be biased because of his EPRI affiliation. Later, Neutra was cross-examined by Ellen Stern Harris, a member of the Consensus Group and the executive director of Fund for the Environment, a Southern California environmental organization.

Harris asked Neutra if he knew of any cases of public health policy being established before full scientific evidence was available. Neutra replied that the most famous example had occurred during the 1840 cholera outbreak in London, when Dr. John Snow, after seeing epidemiological evidence that people drinking water from the Thames downstream from a sewage outlet had a dramatically higher rate of the disease than other people, took the handle off a water pump from which the sick people had been drinking, and thereby eliminated the epidemic. He went on to point out that scientists still do not understand which of the many components of cigarette smoke are responsible for the development of lung cancer and some of the adverse cardiovascular effects that are associated with cigarette smoking.

Toward the end of his stay on the witness stand, Neutra was asked by an attorney for the California Municipal Utilities Association if he still believed that the epidemiological results describing the association between exposure to power-line magnetic fields and cancer had not been confirmed as real.

"I'm in the process of reviewing my views," Neutra replied.

Later that day, John Dawsey, the environmental health administrator for San Diego Gas & Electric, took the witness stand and was asked by an attorney for a ratepayers' group what his position would be if the Commission should find that the relevant research taken as a whole neither demonstrated nor ruled out the possibility that electromagnetic-field exposure could pose a health risk to humans.

Dawsey replied as follows: "Typically, the way the scientific process works is that you establish a hypothesis and then you attempt to demonstrate that that is the case. To date, science has not demonstrated that fields pose a health hazard, nor has science identified any exposure to a component to any element as a possible health risk, and the language was chosen this way and suggested that it be incorporated just to be up front with the public and let them know that this is cautionary, no health hazard has been identified, and to let them know that there's really no basis for representing to them that the public will derive any particular benefit from the implementation of these field management measures, [measures designed to reduce magnetic-field emissions from new power lines] and so to whatever extent money is expended, it's cautionary."

Back in August of 1986, Dawsey's thinking on the power-line problem was somewhat less convoluted. At that time, he had written a report for San Diego Gas & Electric in which he had declared (using words he had obviously borrowed from the Southern California Edison review for which Sahl had been project manager) that although it had not been convincingly demonstrated that occupational or community exposure to power-line magnetic fields was associated with the development of cancer, "the possibility cannot be disregarded and should be seriously considered, given the present data."

A refreshing element of candor was introduced into the hearings on December 10th by Ellen Stern Harris, who had declared in her written testimony that one should not wait for perfect science before developing a public health policy regarding electromagnetic-field exposure. Harris was cross-examined by Gregory Barnes, who asked, "Do you have any idea who will pay for the implementation of the proposals you put forth in your joint testimony?"

"I think that will be largely up to the Public Utilities Commission to determine whether it will be ratepayers or shareholders," Harris replied.

Barnes then asked, "Isn't it a fact, Ms. Harris, that whatever the initial allocation is, that ratepayers will end up bearing most of the burden of any EMF response measures undertaken by the utilities?"

"Probably," Harris replied. "And they will be the same ones who will bear the cost of increased risk to leukemia, brain cancer, learning disabilities, and other things attributed to EMF exposure."

CHAPTER TWENTY-TWO

If People Were Dropping Dead

I N JANUARY of 1993, a motion was made before Judge Galvin to accept the Swedish studies into evidence, but on April 20th he denied it, on the grounds that it was being sought after the conclusion of the hearings, and that a need for including the studies in the record had not been substantiated. Meanwhile, as might be expected, the Swedish studies had engendered considerable consternation among the nation's utility industry and its research arm, the Electric Power Research Institute, which had been insisting for twenty years that there was no real evidence upon which to indict power-line magnetic fields as a health hazard. On October 8, 1992, the Institute issued a bulletin entitled "EPRI Comments on Swedish EMF Residential Study Results," which concentrated on criticizing some of the study's methodology and did not emphasize its results. On November 5th, a report prepared by a consultant for Northeast Utilities declared that a major strength of the Swedish childhood residential study was the ability of its authors to calculate historical exposure to magnetic fields, and that its major limitation was the small number of cancer cases it de-

scribed. The Northeast Utilities review also pointed out that
no lifetime studies of animals exposed to power-line magnetic fields had ever been performed. It made no attempt
to explain why EPRI, which had been in charge of awarding
utility-financed studies of the power-line hazard for two
decades, had never seen fit to award a study of lifetime
animal exposure to power-line magnetic fields.

Not surprisingly, the Swedish studies engendered considerable interest in the nation's press. The October 26th
issue of *Time* ran a piece declaring that they provided "the
best evidence so far of a link between electricity and cancer." The November 8th Sunday edition of the *Los Angeles
Times* published a lengthy article which ran under a headline that read "STUDIES STIR FEARS OVER CANCER RISK
FOR CHILDREN." On the next day, the *Fresno Bee* ran an
abbreviated version of the *Times* piece, together with a box
declaring that the high-voltage transmission lines near the
Slater Elementary School "may have caused the unlikely
number of cancers that have stricken teachers, aides and
students." On November 12th, the *Boston Globe* ran a front-
page story which said that Swedish researchers had reported finding a nearly fourfold increase in leukemia cases
among children exposed to power-line magnetic fields, and
that Swedish government authorities had announced that
they intended to act on the assumption that such exposure
was linked to the development of cancer. A day later, the
San Diego Union-Tribune ran a story that not only described the results of the Swedish studies, but also the findings of a childhood residential study that had been
conducted by Dr. Jorgen H. Olsen, an epidemiologist at the
Cancer Registry of the Danish Cancer Control Agency, in
Copenhagen. According to the *Union-Tribune,* Olsen and
some colleagues had found that Danish children living near

high-voltage transmission lines had a fivefold risk of developing lymphoma. (The Danish study found this fivefold risk among children with average exposure to power-line magnetic fields of one milligauss or more; it also found a 5.6 increased risk for leukemia, brain tumors, and lymphoma combined among children with average magnetic-field exposures of four milligauss or more. In addition, a Danish occupational study found that men working at jobs with chronic magnetic-field exposure of more than three milligauss had a significant sixty-four per cent increased risk of developing leukemia than less-exposed workers.) The *Union-Tribune* article quoted Neutra as saying that the achievement of the Swedish childhood study was "monumental," and that "to be able to go back to 1960 and figure the average exposure of a child is an epidemiologist's dream." When asked whether the Swedish studies showed that magnetic fields caused cancer, Neutra replied that they "certainly move us in that direction. How much more — an inch, 100 yards or a mile — I'm not going to say. I haven't made up my mind yet, but I am getting a clearer picture."

On December 11th, *Science* magazine published a roundup piece on the electromagnetic-field issue by Richard Stone, which ran under a headline declaring that the findings of the Swedish studies had been contradicted by a recently released White House review. "Newspapers and magazines seized on the as yet unpublished Swedish findings as evidence that EMFs pose an invisible threat to health lurking in backyards all across the world," Stone wrote. "But do they? Not according to another recent authoritative source: a report from the Committee on Interagency Radiation Research and Policy Coordination (CIRRPC), part of the White House Office of Science and Technology Policy (OSTP)." Stone went on to quote from the White House review, which had been conducted by an eleven-member panel of experts selected by the Oak Ridge Associated Universities, of Oak Ridge, Tennessee, as stating that "there is no convincing evidence in the published literature to support the contention that exposures to extremely low frequency electric and magnetic fields (ELF-EMF) generated by sources such as household appliances, video display terminals, and local powerlines are demonstrable health hazards."

Later in his article, Stone described Dr. Dimitrios Trichopoulos, chairman of the Department of Epidemiology at Harvard's School of Public Health, who had authored one of the review's chapters on epidemiology, as declaring that even if the Swedish studies had been available for inclusion in the White House report, their findings would not have changed any of its conclusions. (Trichopoulos's fellow panel members backed this assertion in a letter that was subsequently published in *Science*.) Trichopoulos went so far as to tell Stone that if he were to undertake to coordinate a proper study of the power-line magnetic-field problem, "my bet is that it would be a negative study." Nowhere in his article did Stone report that Trichopoulos had testified as a consultant for Crowell & Moring, the electric utility industry's lead attorneys, at the Scientific Advisory Board hearings in the winter of 1991. Nor did he mention that during the autumn of 1990, D. Allan Bromley, President Bush's science adviser, had tried to pressure E.P.A. officials into delaying the release of their draft report describing power-line magnetic fields as a "possible" carcinogen, until it could be reviewed by the Committee on Interagency Radiation Research and Policy Coordination, an offshoot of the White House Office of Science and Technology, which he (Bromley) directed.

The extent to which at least one member of CIRRPC was inclined to dismiss the power-line hazard out of hand was made blatantly obvious by Alvin Young, the committee's chairman, who is the director of the Office of Agricultural Biotechnology at the Department of Agriculture, but who had obviously either not read or chosen not to believe the several dozen studies that had been published in the peer-reviewed medical literature, showing that children and adults exposed to power-frequency magnetic fields at home and at work were developing and dying of leukemia, brain cancer, and lymphoma at significantly higher rates than other people. Around the time the White House review was released, Young told *Microwave News* that research on electromagnetic fields should not receive a high priority or additional funding. "If people were dropping dead from this, it would be different," he declared.

Extraordinary as it may seem, the Swedish residential and occupational studies that had been the impetus for dozens of newspaper articles during the autumn of 1992 and the winter of 1993, as well as segments of several nationally broadcast television programs, went unmentioned during that entire period by the *New York Times,* whose coverage of the power-line health hazard over the years has been spotty at best and often non-existent. (Interestingly, on Sunday, May 2nd, 1993, the *Times* saw fit to run an Associated Press story about the outcome of the San Diego power-line trial, under a misleading headline that read "JURY REJECTS THEORY THAT POWER LINES CAUSED GIRL'S TUMORS." In fact, the jury had made no such finding. It had merely found that San Diego Gas & Electric had not been negligent for not warning back in 1986 that power-line magnetic fields might pose a health hazard.) Nor did the Swedish studies receive anything more than passing mention in the *Washington Post,* whose health editor had been quoted in the January/February 1991 issue of the *Washington Journalism Review* as declaring that the electromagnetic-field issue has been "hyped" and "doesn't deserve coverage at all." Nor were the findings of the Swedish studies mentioned in the *Wall Street Journal* until February 5, 1993, when the newspaper devoted twenty-two words to them in a twenty-five-hundred-word front-page story, which ran under a headline that read "ELECTRIC UTILITIES BRACE FOR CANCER LAWSUITS THOUGH RISK IS UNCLEAR." The *Journal* story went on to quote Robert Adair, of Yale, as complaining that concern about the Swedish studies and the power-line hazard was nothing more than "electrophobia."

The reluctance of three of the nation's leading newspapers to consider the Swedish studies newsworthy notwithstanding, four thousand five hundred and sixty-seven readers of an article on the electromagnetic-field hazard that appeared in the January 1–3, 1993, issue of *USA Weekend* — a Sunday supplement magazine with a readership of thirty-three and a half million — showed the importance they attached to the issue by filling out a form entitled "MEMO TO: AL GORE," in which they selected what they considered to be the nation's number one environmental health priority. According to the article, E.P.A. officials who tested Gore's Senate office for electromagnetic fields in 1991 had found background levels of nine milligauss, and Gore had then borrowed their gaussmeter to test his home. *USA Weekend* undertook to deliver the four thousand five hundred and sixty-seven responses to its poll to Gore during inauguration week. The results of the poll were announced in the February 19–21 issue of the supplement.

Thirty-five per cent of the readers who filled out the memo form selected electromagnetic fields as the nation's number one environmental health priority; seventeen per cent selected chemicals in food; twelve per cent chose indoor air pollution; and thirty-six per cent listed other environmental concerns.

Meanwhile, teachers and parents at schools located near power lines in various parts of the nation were expressing alarm over the results of the recent Swedish and Danish studies. On November 16, 1992, the electromagnetic-field task force at the Montecito Union School — it had been largely inactive during the two years since Neutra and his colleagues in the Department of Health Services had completed their investigation of the unusual cluster of leukemia and lymphoma among children attending the school — met to discuss the Scandinavian findings and to consider whether any new measures should be taken to protect teachers and students at the school. (Back in 1990, the members of the task force had allowed themselves to be persuaded that two milligauss was a safe level of exposure for children, and they were understandably worried now that the Swedish study had found twice the expected risk of leukemia, and the Danish study had found five times the expected risk of lymphoma, in children exposed to power-line magnetic fields of only one milligauss or more.) In an article that appeared in the *Santa Barbara News-Press* on the day of the meeting, Melinda Burns reported that just a few days earlier, when Imre Gyuk, the program manager of electromagnetic research at the U.S. Department of Energy, had been shown a map of Montecito Union and its adjacent power lines, together with measurements of their magnetic-field emissions, he had said, "If I were a parent, I would get the hell out." According to the newspaper, Gyuk tried to retract this statement on the following day, telling Burns that he would be "very much concerned," and would "certainly be active in the PTA."

On January 12, 1993, members of the Montecito task force recommended unanimously that a medical survey going back to 1950 be conducted of teachers, teacher's aides, and administrative staff members at the school, and that additional measurements of electromagnetic fields be taken on the school grounds. They also recommended that Southern California Edison consider moving its switching and voltage-transforming operations from the substation next to the school to a substation in nearby Summerland that is not situated near any school or dwellings.

In an interview with Melinda Burns a day later, Jack Sahl said that it was too early to judge the significance of the Swedish studies, and that state and federal health officials simply did not know if electromagnetic fields could cause cancer. "It's not like industry is fighting your health departments," Sahl said. "We are funding research to find the answer. Science is an extraordinarily long process that takes a lot of time. We don't know what the exposure is that we're supposed to be concerned about."

Indication that some of the parents of children attending Montecito Union were fed up with this kind of response came on February 4th, when Sam Tyler, a parent representing an ad hoc group of eight or so other parents, told the Montecito School Board that the power lines and the substation must be removed to restore the integrity and spirit of the school. In an article that appeared in the *News-Press* on February 5th, Burns quoted him as saying, "The substation just shouldn't be here. It is going to become symbolic in our community. This issue is clearly catching on. People will not buy houses in Montecito because of the substation. Until it is removed, no matter what you do, [this issue] will continue to heat up."

While parents and school officials were grappling with the power-line problem at Montecito Union, a situation similar to the one that had occurred at the Slater Elementary School in Fresno was unfolding at the Montague Elementary School, in Santa Clara — a city of about ninety thousand inhabitants that is located in the South Bay area, about thirty-five miles southeast of San Francisco. During the autumn of 1992, Marilyn Pope, the mother of a fifth-grader at the Montague School, had asked that magnetic-field measurements be taken at the school, whose south side sits within eighty feet of a one-hundred-and-fifteen-thousand-volt transmission line. At the request of officials of the Santa Clara Unified School District, engineers from Pacific Gas & Electric, owner of the transmission line, came to the school on December 2nd and measured fields that ranged from 3.7 to 5.3 milligauss in the centers of four classrooms nearest the line. Magnetic-field levels in the centers of four adjacent classrooms ranged from 2 to 3.1 milligauss. Field levels in the centers of eight classrooms located in a part of the school that is farther from the transmission line ranged from 0.9 to 2.2 milligauss. Levels as high as 8.7 milligauss were found on the school playground.

At a meeting of the Santa Clara Board of Education, which was held on January 7, 1993, Pope said that either children should be moved from the classrooms nearest the transmission line, or P. G. & E. should bury the line. According to Pope, P. G. & E. officials who attended the meeting claimed that burying the line would simply bring it closer to the school, and they professed to be ignorant of the 1989 study coordinated by the Empire State Electric Energy Research Corporation, which showed that when high-voltage transmission lines are buried correctly, the magnetic fields they emit can be sharply reduced.

Pope's concern about the power-line hazard was questioned by Dr. Wallace Sampson, who is chief of oncology at the Valley Medical Center in Santa Clara. According to an account that appeared in the next day's *San Jose Mercury News,* Sampson said that he knew of no convincing evidence linking electromagnetic fields and cancer. "As an overall health problem, it pales in comparison to alcohol, tobacco and other self-destructive behavior," Sampson declared. He appeared not to appreciate the fact that the exposure of teachers and children at the Montague School to power-line emissions was involuntary.

On January 21st, Patricia Duran, the widow of Robert Duran, who had died of cancer three years earlier, after teaching for ten years in one of the classrooms closest to the transmission line, spoke at a meeting of the Santa Clara Unified School District. She told the trustees of the district that if there was something wrong at the school, it needed to be shut down and the children needed to be moved out. At the same meeting, school district officials announced that they had decided to move students and teachers from the four classrooms nearest the transmission line to the school library. The decision of the school officials was prompted in part by demands set forth by teachers at the school, who claim that there has been a high incidence of cancer and at least three cancer-related deaths among them during the past ten years. (There are about fifteen teachers and three hundred and fifty students at Montague Elementary.) According to an article published in the April, 1993, issue of *CTA Action,* a monthly newsletter put out by the California Teachers Association, thirteen of the fifteen teachers at Montague Elementary had requested transfer to other schools. The transfer requests, which were filed on March 1st, preserved the teachers' contractual right to change schools if corrective action was not taken within the next five months. "P. G. & E. has got to do something to make our school safe," the newsletter quoted one of the teachers as saying. She and her colleagues at the Montague School were given full support by the United Teachers of Santa Clara, which has offered to act as a clearinghouse for information about power lines and health problems in California schools. This, of course, is something that Neutra and the State Department of Health Services could have undertaken in conjunction with the California Department of Education several years earlier, when there was good reason to believe that the cluster of leukemia and lymphoma at Montecito Union may have been related to power-line emissions.

CHAPTER TWENTY-THREE

The Writing on the Wall

I N VIEW of the growing body of evidence that exposure to power-line magnetic fields can cause or promote cancer, especially in children, further delay in the implementation of preventive measures to protect the health of schoolchildren exposed to these fields seems stupid at best and criminally stupid at worst. At the Tamalpais Valley Elementary School, in Mill Valley — a town of about thirteen thousand inhabitants on Marin Peninsula, just north of San Francisco — school district officials have wisely decided to vacate some classrooms that are located close to a high-voltage transmission line, and they have been negotiating with Pacific Gas & Electric, owner of the line, about raising, moving, or re-phasing the line, in order to reduce the magnetic-field exposure of children and teachers at the school. Elsewhere in Mill Valley, a primary distribution line running next to the Park Elementary School on East Blithedale Avenue is putting fields of up to four milligauss in classrooms that are located in the wing of the school nearest the line. According to an article by Peter White that appeared on March 14th in IMAGE — a Sunday magazine that is pub-

lished by the *San Francisco Examiner-Chronicle* — P. G. &
E. officials have said that reconfiguring the wires might
partly reduce the strength of the magnetic fields at Park Elementary, but that if undergrounding them proved necessary, the cost for doing so in the one block next to the
school would come to $150,000. White reported that at a
January 14th meeting with parents and school officials, Mar-
tha McNeal, P. G. & E.'s program director for electromagnetic fields, said that the utility would be willing to spend
money to reduce magnetic-field emissions from new power
lines and electrical facilities, but that the company was not
willing to spend significant amounts of money to reduce
the fields given off by existing power lines, and, therefore,
would not take action to lower the levels at the Park School.
When Sonya Shaw, the mother of two children who attend
the school, asked McNeal if she thought there was a problem with the electromagnetic fields that had been measured there, McNeal replied, "Our perception is that you
perceive there is a problem."

Since this meeting, P. G. & E. appears to have modified
its position, and has begun working with parents and
school district officials in an attempt to achieve solutions to
the magnetic-field problems that exist at both Tamalpais Valley and Park elementary schools. This does not mean that
the utility has decided that the time has come for it to
honor the pledge it made in September of 1991, when it
mailed a brochure to tens of thousands of customers, assuring them that "if scientific investigations demonstrate a
cause-and-effect relationship between EMFs and adverse
health impacts, the company will work with governmental
agencies to take appropriate measures to address health
risks to employees and the public." Rather, the reason for
P. G. & E.'s apparent change of heart may be found in some

events that took place at the Slater Elementary School, in Fresno, during the winter of 1993.

On February 2nd, Neutra sent a letter to Pat Berryman and her colleagues on the Slater Electromagnetic Field Study Subcommittee, telling them that the Department of Health Services "does not believe that the weight of evidence linking EMF to cancer is strong enough to warrant mandating the closure of the estimated 1000 schools in California next to power lines." (Not only did he fail once again to provide any substantiation whatsoever for the claim that one thousand schools in California were close enough to high-voltage or high-current power lines to warrant closure, but he chose to ignore the fact that school officials at the Slater, Montague, and Tamalpais Valley elementary schools had not closed down entire schools, but had simply undertaken to shut down some classrooms in their schools that were situated next to power lines giving off strong magnetic fields.) Later in his letter, Neutra told the subcommittee members that "unless and until this Department feels that the weight of evidence is sufficient to require potentially costly regulations, local jurisdictions have to find the best solution to these situations." He also declared that "as a matter of policy in 1991 and now, we cannot justify health based recommendations for the school, either to move the teachers back to Pods A or B or to maintain the closure of that wing."

Neutra had been scheduled for several weeks to present his final report in person to the members of the Slater study subcommittee, on February 24th. However, at the outset of the meeting, school district officials informed Berryman and other subcommittee members that Neutra would not be coming. They went on to say that since they understood that his findings would be inconclusive, they had decided to make the evacuation of Pods A and B permanent, and to use the classrooms in those pods for storage rooms. The school officials pointed out that this action would, in effect, bring the Slater School within the California Department of Education's requirement that all new schools be at least a hundred and fifty feet from high-voltage transmission lines. Additionally interesting was what they did not see fit to tell the subcommittee members.

Almost a month earlier, the *Fresno Bee* had reported that the son of Katie Alexander — the first-grade teacher who had died of brain cancer in April of 1991, after having taught in Pod A for fifteen years — had filed a wrongful death suit against P. G. & E. in Fresno Superior Court, claiming that the transmission lines next to the school had caused his mother's death. The newspaper also said that family members of other Slater teachers were contemplating legal action. Two days later, Evelyn Hurd, the widow of Curtis Hurd — the vice-principal who had died of colon cancer in February of 1992 — filed a wrongful death suit against the utility. What the school district officials neglected to tell the subcommittee members was that on January 15th, Evelyn Hurd had filed a worker's compensation claim against the school district, alleging that her husband's death had been caused by his having worked in electromagnetic fields created by the high-voltage transmission lines on Emerson Avenue, and that the school district had not provided him with a safe place of work. They also neglected to mention that the district's worker's compensation insurance company had taken the highly unusual step of filing as an intervenor against P. G. & E. in the wrongful death suit brought by Evelyn Hurd, even before it had been required to pay out any money, in order to be in a position to recoup from the utility whatever costs it might incur should Hurd's widow be successful in her worker's compensation claim. Thus, the school district officials had an excellent legal reason for disregarding the inconclusivity of Neutra's final report, and for announcing that they were permanently closing Pods A and B. Indeed, they had probably been advised to do so by their attorneys.

At about this time, Berryman and her colleagues learned that George Marsh, who had been the principal at the Slater School in the winter of 1991, when teachers there first began to suspect that power-line emissions might be causing the unusual incidence of cancer among them, had developed prostate cancer, a malignancy that was found to be elevated in the telephone-company cable splicers who had been surveyed by Matanoski. Marsh, whose office had been located in the administrative area directly behind Hurd's, thus became the fifteenth person at Slater to have developed cancer after working on the side of the school nearest the transmission lines on Emerson Avenue.

On April 10th, the *Fresno Bee* finally got around to running a story about the Department of Health Services' draft report of October 15, 1992. It appeared under a banner headline that read "SLATER CANCER, POWER LINES AREN'T CONNECTED, STATE SAYS." The newspaper quoted Neutra and his colleagues as declaring that "at this point, we can't say a definitive yes or a definitive no" as to whether electromagnetic fields emanating from two power lines caused cancer in school employees and students. The *Bee* described Berryman as saying that state officials had conducted no measurements at the Slater School, and had made no effort to track down former Slater teachers and students, but had relied on teachers to take measurements and to track down former colleagues and students. "None of that should have been done by us," Berryman told the

Bee. "It should have been done by the state of California."

The article in the *Bee* said that a twenty-six-year-old Fresno man who attended the Slater School from kindergarten to the sixth grade had developed brain cancer, and that a nine-year-old girl who attended first and second grade had recently been diagnosed with a rare bone cancer. If the newspaper had inquired where these former students had lived, or had canvassed houses along Emerson Avenue to determine whether there might be an unusual incidence of cancer among people living in them, it could have learned the following: the twenty-six-year-old man with brain cancer had lived most of his life on Emerson Avenue, a short distance from the Slater School, and about a hundred feet from the one-hundred-and-fifteen-thousand-volt transmission line; a young woman who has been diagnosed with a nonmalignant brain tumor, and whose bedroom is within forty feet of the two-hundred-and-thirty-thousand-volt line, lives on the other side of Emerson Avenue within sight of the school; and a woman who lived on Emerson Avenue in a house situated about seventy feet from the two-hundred-and-thirty-thousand-volt line, and about a quarter of a mile from Slater Elementary, had died a year or so ago from a brain tumor. If the *Bee* had undertaken such an investigation — it had, after all, been covering the cancer cluster at the Slater School for well over two years — it might have been in a position to question why Neutra and his colleagues in the Department of Health Services had not done so, and to have pointed out that when faced with a similar situation three years earlier at the Montecito Union School, their response had been similarly deficient.

The article in the *Bee* dutifully described the appendix that Neutra had attached to his October draft report as estimating that about one thousand out of California's eight thousand schools are situated near power lines, and that fifty-six of these one thousand schools could be expected to have eleven or more cases of cancer "by chance alone." The appendix was, of course, incomplete and out of date. It did not include two cases of cancer in employees who had declined to participate in the state's study, or the colon cancer that had been diagnosed during the autumn of 1992 in the Pod B teacher, or George Marsh's prostate cancer, which had only recently been discovered. But it was not these omissions that made Neutra's report ring hollow; rather, it was his insistence on ignoring the true meaning of the fact that *all* of the cancers that had been diagnosed among teachers, teacher's aides, and staff members at the Slater School had occurred among people who worked on the side of the school nearest the transmission lines. In doing so, Neutra deliberately chose to ignore the essential health issue at the school, just as in his February letter to the members of the Slater Electromagnetic Field Study Sub- committee, he had chosen to ignore the essential issue of the situation in California schools as a whole, when he in- voked the dire prospect of closing down one thousand schools in the state, while disregarding the unlikelihood of any school being surrounded by power lines, and the fact that the only closings that had occurred had not been of en- tire schools, but of classrooms on the sides of schools that were next to power lines. The inadequacy of Neutra's as- sessment of the true nature and extent of the cancer hazard at the Slater School can easily be seen, when, assuming the null hypothesis that a cancer case is equally likely to be found in someone working on either side of the school, one undertakes to test the probability that fifteen cancer cases would be found among people working on the side nearest the power lines, and that no cancers would be found among people working on the other side.

A simple sign test will show that the probability of that occurring is considerably less than fifty-six out of one thousand. Indeed, it is one out of 2^{15}, or one in 32,768.

In spite of steadily mounting evidence that power-line magnetic fields can cause or promote cancer, most of the nation's utilities are continuing to insist that proof of any connection is lacking and that further research is necessary before preventive measures can be implemented, even at schools with high magnetic-field levels. However, some cracks have begun to appear in the utility industry's stone wall. In Brentwood, a wealthy suburb of Santa Monica, the Kenter Canyon Elementary School sits next to a right-of-way for a two-hundred-and-thirty-thousand-volt transmission line and two one-hundred-and-thirty-eight-thousand-volt lines, which carry power for the entire West Los Angeles area. When magnetic-field levels of nearly twelve milligauss were measured on the grounds of the school in the spring of 1992, school authorities and parents brought pressure to bear on the Los Angeles Department of Water and Power, owner of the transmission lines, to do something about the problem. Since then, the Department has re-phased the lines, which has reduced their magnetic-field emissions by fifty per cent, and has announced that it is planning to take measures that will reduce magnetic-field levels even further.

A further crack in the wall occurred in late February, when New York State Attorney General Robert Abrams sent a letter to the chief executive officers of seven of the state's power companies — Rochester Gas and Electric, the Long Island Lighting Company, New York State Electric and Gas,

Central Hudson Gas and Electric, Consolidated Edison, Orange and Rockland Utilities, and the New York Power Authority — urging them to "initiate a statewide survey and electromagnetic field measurement program of all schools, primary and secondary, that have transmission lines (designed or operated at 69kV and above) on or adjacent to school property." Abrams told the utility executives that Niagara Mohawk, the second-largest utility in New York, had already conducted such a study at his request. In a press release he said, "Rather than bury our heads in the sand and assume a problem does not exist, utilities should compile data on this issue at once."

On March 23, 1993, Abrams sent a press release to the *New York Times* and other leading newspapers, announcing that he had obtained a voluntary agreement from each of New York's eight utilities to undertake a comprehensive survey of power lines near schools, as well as to measure the strength of the magnetic fields these lines were emitting. He also announced that Niagara Mohawk had taken significant steps to reduce electromagnetic fields at the Voorheesville School, in Albany County. According to Abrams, Niagara Mohawk had substantially reduced the fields emanating from a power line that runs within ten feet of the school by decreasing its current load, and had promised to take the line out of service by September 1st. The utility also said that it would reconfigure the wires of a second power line, which is located within seventy feet of the school, so that the magnetic fields it emits will be reduced. What was remarkable about Niagara Mohawk's action is that it came just a few months after the utility had sent out a color brochure to its customers, which played down the childhood cancer risk posed by magnetic-field exposure, and pointed out that daily activities, such as automobile driving, were "much more hazardous than EMF exposure."

No mention of Abrams's letter or the reaction of New York State's eight utilities appeared in the *New York Times*. Nor did the *Times* see fit to print the fact that in April the New Jersey Board of Regulatory Commissioners requested utilities in that state to conduct a survey similar to the one that had been conducted in New York, and to measure the electromagnetic-field levels at schools situated within a hundred feet of any transmission line carrying sixty-nine thousand volts or more. Meanwhile, the *New York Observer* had reported that between 1990 and 1992, two students at the Town School — a private elementary school on East 76th Street, in Manhattan, which is located next to a Consolidated Edison substation — had developed leukemia. (One of the students had died, and the other — a third-grader — is undergoing treatment.) In January of 1991, magnetic-field levels of about ten milligauss were measured in the kindergarten of the Town School, and during the spring school officials moved the kindergarten to another part of the building, and undertook to shield a high-voltage power line that runs through a wall abutting a neighboring apartment building. According to the *Observer,* a spokesperson for Con Edison said that scientists had not been able to determine the biological mechanism by which electromagnetic fields can affect people.

By the spring of 1993, the claim that no generally accepted mechanism to explain how magnetic fields could interact with cells to cause cancer was fast becoming the utility industry's answer to the Swedish studies, and its fallback position on the power-line health issue. What industry spokespeople conveniently overlooked, of course, was that thirty years after definitive epidemiology had been conducted to show that asbestos was a potent cancer-producing agent, scientists still do not know the mechanism by which an inhaled asbestos fiber reacts in lung tissue to cause cancer. Nor do they understand the mechanism by which cigarette smoke reacts in lung tissue to cause cancer. Or how the chemical pesticide DDT operates in breast tissue to cause breast cancer. Suffice it to say, if public health authorities had been required to wait for the cancer-producing mechanisms of these agents to be fully understood, regulations governing asbestos exposure would not have been implemented; warnings on cigarette smoking would not have been issued; and the twenty-year-old ban on DDT would not have been imposed.

As it happens, however, recent laboratory experiments showing that extra-low-frequency magnetic fields can greatly enhance the production of Insulin-like Growth Factor-II — a growth hormone that under normal conditions is known to play a role in human growth and development — have provided what may well prove to be a giant step toward understanding how magnetic fields can promote the development of cancer, because unregulated enhancement of IGF-II has been implicated in the development of brain cancer, kidney cancer, and a number of other malignancies.

What seems clear at this point is that most of the nation's utilities will go on proclaiming that "the jury is still out" on the issue of whether power-line magnetic fields are linked with cancer, until real jurors who have assessed the epidemiological and experimental evidence have found either for or against the proposition that a preponderance of the credible evidence shows that the fields do, in fact, cause or promote cancer. In the meantime, a defense claiming that the medical and scientific evidence is inconclusive may well prevail in some of the early lawsuits, just as a similar defense prevailed twenty years ago in some of the early asbestos trials. The utilities would do well to remember, however, that the defense of inconclusivity proved to be a sand castle in the path of a rising tide of proof that asbestos was extremely dangerous to inhale, and that such a defense may well prove to be equally frail if evidence showing that power-line magnetic-field exposure is hazardous continues to pile up in the future as it has in the past.

The fact is that the utilities are staring at some ominous writing on the wall in the form of massive citizen action that is being mounted from one end of the nation to the other to prevent the construction of new voltage and high-current power lines in residential neighborhoods, and to demand that existing lines be reconfigured, rerouted, or buried to reduce magnetic-field exposure. Indeed, it appears that the utilities may have made a grave miscalculation in believing that they could persuade the American people that the presumption of innocence should be extended to power-line magnetic fields while further studies of the hazard are conducted over yet another decade. In any event, it is quite obvious that the utility industry and state and federal public health officials are rapidly losing control of the situation regarding the power-line hazard, which is now passing by default to the people, who are giving stronger and stronger indication with each passing week that they will not tolerate the continuing exposure of themselves and their children to power-line magnetic-field levels that have already been associated with the development of cancer in a majority of the sixty-odd epidemiological studies that have so far been conducted.

If things continue on their present course, it seems safe to predict that the American people will one day conclude that it is not necessary to have incontrovertible evidence that power-line emissions are hazardous to human health before taking preventive measures to reduce exposure, any more than it was necessary to have incontrovertible evidence that cigarette smoking causes lung cancer before warning the public about it. It also seems reasonable to predict that the people will then demand that their government acknowledge the existence of the power-line hazard, just as the Swedish Government has done. How soon this takes place will be partly determined by the outcome of the legal warfare that has just begun, and by how much public pressure is brought to bear on state and federal health officials, who continue to consider the power-line hazard as a political and economic problem, instead of as a medical problem. At this point, a likely long-term scenario would seem to include mounting citizen action and protest, proliferating lawsuits, jury verdicts sooner or later for the plaintiffs, out-of-court settlements, spiraling legal costs, and, ultimately (because the public will be asked to foot the bill for all of this), public outrage with profound political repercussions and regulatory consequences for how the electric utility industry will be allowed to conduct its business in the future.

Unfortunately, the power-line cover-up will have additional consequences: thousands of unsuspecting children and adults will be stricken with cancer, and many of them will die unnecessarily early deaths, as a result of their exposure to power-line magnetic fields.

INDEX